SEP 2 7 1954

HISTORICAL RECORDS AND STUDIES

THE MONOGRAPH SERIES

I. The Voyages of Christopher Columbus, as Told by the Discoverer
II. Unpublished Letters of Charles Carroll of Carrollton and of His Father, Charles Carroll of Doughoregan
III. Forty Years in the United States of America (1839-1885) By the *Rev. Augustus J. Thébaud, S.J.*
IV. Historical Sketch of St. Joseph's Provincial Seminary, Troy, New York. By the *Right Rev. Henry Gabriels, D.D.*
V. The Cosmographiæ Introductio of Martin Waldseemüller In Facsimile
VI. Three Quarters of a Century (1807-1882) By the *Rev. Augustus J. Thébaud, S.J.* Two Vols.
VII. Diary of a Visit to the United States of America in the Year 1883. By *Charles Lord Russell of Killowen*, late Lord Chief Justice of England
VIII. St. Joseph's Seminary, Dunwoodie, New York, 1896-1921 By the *Right Rev. Arthur J. Scanlan, S.T.D.*
IX. The Catholic Church in Virginia (1815-1822) By the *Right Rev. Peter Guilday, Ph.D.*
X. The Life of the Right Rev. John Baptist Mary David (1761-1841). By *Sister Columba Fox, M.A.*
XI. The Doctrina Breve (Mexico 1544) In Facsimile
XII. Pioneer Catholic Journalism By *Rev. Paul J. Foik, C.S.C., Ph.D.*
XIII. Dominicans in Early Florida By the *Rev. V. F. O'Daniel, O.P., S.T.M., Litt.D.*
XIV. Pioneer German Catholics in the American Colonies (1734-1784). By the *Rev. Lambert Schrott, O.S.B.*—The Leopoldine Foundation and the Church in the United States (1829-1839). By the *Rev. Theodore Romer, O.M.Cap., S.T.B., M.A.*
XV. Gonzalo de Tapia (1561-1594), founder of the first permanent Jesuit Mission in North America, by *Rev. W. Eugene Shiels, S.J., Ph.D.*
XVI. Old St. Peter's. The Mother Church of Catholic New York (1785-1935). By *Leo Raymond Ryan, A.B., M.S. (E.)*
XVII. The Quebec Act: A Primary Cause of the American Revolution. By *Rev. Charles H. Metzger, S.J.*
XVIII. Catholic Immigration Colonization Projects in the United States, 1815-1860. By *Sister Mary Gilbert Kelly, O.P., Ph.D.*
XIX. Adventures of Alonso: Containing Some Striking Anecdotes of the present Prime Minister of Portugal; published in London, 1775. By *Thomas Atwood Digges.*
XX. Thomas Francis Meehan (1854-1942). A Memoir By *Sister M. Natalena Farrelly, S.S.J.*
XXI. El Rio del Espíritu Santo. By *Rev. Jean Delanglez, S.J., Ph.D.*

UNITED STATES CATHOLIC HISTORICAL SOCIETY

Honorary President
MOST REVEREND FRANCIS J. SPELLMAN, D.D.

President
ARTHUR F. J. RÉMY, PH.D.

Vice-President
MONSIGNOR PHILIP J. FURLONG, PH.D.

Treasurer
CHARLES H. RIDDER

Corresponding Secretary
LEO R. RYAN, PH.D.

Recording Secretary
GEORGE B. FARGIS

Executive Secretary
ELIZABETH P. HERBERMANN

Editor of Publications
MONSIGNOR THOMAS J. MCMAHON, S.T.D.

Librarian-Archivist
THOMAS F. O'CONNOR, M.A.

Trustees
RT. REV. MGR. A. J. SCANLAN, S.T.D.
VERY REV. THOMAS J. MCMAHON, S.T.D.
WILLIAM THOMAS WALSH
ARTHUR KENEDY
WILLIAM J. AMEND
REV. EDWARD J. KERN, PH.D.
RICHARD REID
REV. FRANCIS X. TALBOT, S.J.

Councillors

JOSEPH C. DRISCOLL
JOSEPH H. MCGUIRE
REV. JOHN K. SHARP
CHARLES CALLAN TANSILL

JOHN J. FALAHEE
S. STERNS CUNNINGHAM
REV. W. EUGENE SHIELS, S.J.
JOHN J. O'CONNOR

UNITED STATES CATHOLIC HISTORICAL SOCIETY

MONOGRAPH SERIES XXI

EL RIO DEL ESPÍRITU SANTO

An Essay on the Cartography
of the
Gulf Coast and the Adjacent Territory
during the
Sixteenth and Seventeenth Centuries

by
JEAN DELANGLEZ, S.J., PH.D.

Edited by
THOMAS J. McMAHON, S.T.D.

COPYRIGHT 1945
THE UNITED STATES CATHOLIC
HISTORICAL SOCIETY

Copyright 1945
THE UNITED STATES CATHOLIC
HISTORICAL SOCIETY

PUBLICATION COMMITTEE
Very Rev. Thomas J. McMahon, S.T.D.

S. Sterns Cunningham   Richard Reid

Rev. W. Eugene Shiels, S.J., Ph.D.

OFFICE OF
THE EXECUTIVE SECRETARY OF THE SOCIETY
SUITE 103
924 WEST END AVENUE, NEW YORK 25, NEW YORK

FOREWORD

In presenting the present work as the twenty-first in the series of monographs published by the United States Catholic Historical Society, it is hardly necessary to remind our many members and friends that the name of its author is already well known to all devotees of the history of the Church in North America. Dr. Delanglez has been Research Professor of History at Loyola University, Chicago, for the past nine years, and he is recognized as an authority on the part the French missionaries and explorers played in the making of America. Amply equipped for scientific research by a brilliant course in Munich and at the Catholic University of America, which culminated in his doctoral dissertation, "The French Jesuits in Lower Louisiana (1700-1763)", he is the author of many works on analogous subjects. At present he is occupied in editing the late Father Garraghan's "Guide to Historical Method", which is probably the work most desired by students at the present time.

Not the least important among Dr. Delanglez' valued contributions have been the numerous articles on the cartography and the early history of the Great Lakes Region and the Mississippi Valley. From July 1943 through July 1944 there appeared in *Mid-America* much of the contents of the present monograph. Since these articles were in serial form and, therefore, without guarantee of permanence, their truly great value attracted the attention of the Board of Publications of the Society, which decided to include them among the monographs. The author has made many important modifications and additions so as to further emphasize what may well be the definitive thesis on "El Rio del Espíritu Santo".

Thus the Society continues the policy which it has faithfully pursued since 1884, the year of its foundation. Along with these twenty-one monographs, thirty-four volumes of *Historical Records and Studies* are complete justification for its existence among the learned historical societies of the nation. It boasts a large and loyal enrollment of educational institutions, professors and laity throughout the United States and even abroad, and it enjoys the munificent patronage of its Honorary President, Archbishop Spellman. The war has not impeded its work of publication nor has it dimmed the vision and high purpose of the officers of the Society, now looking forward to a campaign for even more members and always planning for greater service to the cause of Mother Church.

THOMAS J. MCMAHON

New York, September 8, 1945

TABLE OF CONTENTS

Foreword vii

Introduction xi

    I. La Bahía del Espíritu Santo 1

    II. The First Group of Maps—The Padrón 11

    III. The Second Group of Maps 33

    IV. The Narrative of Alvar Núñez Cabeza de Vaca 47

    V. The Narratives of the De Soto Expedition—The Third Group of Maps 55

    VI. The Fourth Group of Maps—The Sanson-Jaillot Interpretation 81

    VII. The Discovery of the Mississippi 96

    VIII. The Re-Discovery of the Gulf Coast by the Spaniards 113

    IX. The Coming of the French 134

    X. Conclusion 145

Bibliography 147

Plates 162

List of Maps 162

Index 173

INTRODUCTION

IN MAY 1673, Louis Jolliet accompanied by Father Jacques Marquette and five French-Canadians left Michilimackinac (St. Ignace, Michigan) to search for a water route to the Sea of the South by way of the great river—called Mississippi by the Indians—which, it was believed, discharged itself into the Sea of California. The party ascended the Fox River to Portage, where they crossed over to the Wisconsin, and went down this river to Prairie du Chien. They then descended the Mississippi to near the mouth of the Arkansas River, whence they returned to Green Bay *via* the Illinois and the Des Plaines rivers, through Chicago and up the west shore of Lake Michigan.

The importance of this voyage and the advance in knowledge of the geography of the United States which it effected are better appreciated, if one ascertains what was previously known in Europe and in America of the country traversed by the expedition and of the river which they followed.

This can be done best by studying the maps of the region which were in existence in the sixties of the seventeenth century. The printed accounts of the early voyages along the Gulf Coast and into the interior are few and unsatisfactory. Even the more detailed accounts of the expedition of De Soto give us little that is definite with regard to the geography of the Southern United States. The mapmakers of the time, however, had access to information derived from narratives which are lost today. Unfortunately, such information was not always made use of in the sixteenth and seventeenth centuries; maps of the New World, and especially maps of the interior, were seldom made from accounts of discovery and exploration.

The majority of mapmakers were satisfied with reproducing the work of their predecessors. True geographers were few and far between, and many who are today classed as cartographers were simply draughtsmen, engravers, or map publishers, whose main concern was to issue "new" maps. The only change often consisted in scratching the date of publication off the plate and replacing it by a more recent date, notwithstanding the emblazoned assertion that the map embodied the latest information derived from memoirs of travelers recently returned from the faraway countries represented on the map. Sometimes the map publisher did not even bother to alter the plate. He merely erased the last two digits on the unsold maps in stock, inserted the more recent date with pen and ink, and advertised the map as a new edition. This was not all. Map publishers indiscriminately compiled atlases of dated and undated maps by various authors, with the result that today we find these maps listed under the date of publication of the atlas, which in some cases varies as much as fifty years from the true date of an individual map. It is therefore very difficult, unless one has independent knowledge of the maps in these atlases, to study the cartographical development of a particular area of the New World, such as the Mississippi Valley and the adjacent States with which we are here concerned.

Another convenient way of making a new map, quite different from those already on the market, and hence likely to interest the publisher's clients, was as follows: The cartographer or the draughtsman would insert, more or less indiscriminately, various place-names, rivers, and other geographical features to be found on whatever previously printed or manuscript maps he happened to have at hand. The cartographical freaks produced by this desire for completeness and novelty can easily be imagined.

In the present essay, relying on the numerous maps of the interior of the United States between the Appalachian Range and the Rocky Mountains, we shall inquire what knowledge the best informed geographers of the period seem to have had of the Mississippi, and in particular, whether the river which is labeled "Rio del Espíritu Santo" on these maps is the Mississippi itself. Most writers who have treated the question have either answered the question affirmatively or have taken an affirmative answer for granted without any investigation. Half a century ago, however, on the occasion of the fourth centenary of the discovery of America, the question whether the Rio del Espíritu Santo of the Spanish geographers was the Mississippi was answered negatively by Scaife.[1] Although we shall agree with some of his arguments, his general conclusion cannot be accepted, for he held that the name Rio del Espíritu Santo "applied generally if not exclusively, to the stream which now bears in its different parts, the names of Coosa, Alabama, and Mobile."[2] The main argument

---

[1] W. B. Scaife, *America, Its Geographical History, 1492-1892. . . . With a supplement entitled: Was the Rio del Espiritu Santo of the Spanish Geographers the Mississippi?* Baltimore, 1892, 139-176.

[2] *Ibid.*, 139.

that led Scaife to this conclusion is that on all the earliest maps, the Rio del Espíritu Santo empties directly not into the Gulf, but into a large bay which he identified as Mobile Bay.

We may begin by noting three facts about the Mississippi itself which are beyond dispute. First, within thirty years after the discovery of America, some Spaniards undoubtedly sailed by its mouth without realizing what they were passing. Secondly, in 1541 De Soto saw and crossed the Mississippi somewhere between the Arkansas and the St. Francis rivers, and two years later the remnants of his expedition led by Moscoso descended it to the sea. Finally, long before the era of exploration, the muddy waters of the Mississippi emptied directly into the Gulf just as they do today, without passing through any such bay as appears on early maps at the mouth of the Rio del Espíritu Santo.

EL RIO DEL ESPÍRITU SANTO

I
LA BAHÍA DEL ESPÍRITU SANTO

BEFORE we begin to study the maps themselves, another general observation should be made. On the earlier maps the name Bahía del Espíritu Santo is given to a bay at the northwestern end of the Gulf, whereas at a later date another Bahía del Espíritu Santo appears on the west coast of the Florida peninsula; and a Rio del Espíritu Santo is represented as emptying into each of these bays. To settle which of the two rivers is the Mississippi presents no difficulty, for all those who identify the Mississippi with the Rio del Espíritu Santo hold that it is the western one; however, a study of the changes in the nomenclature of both rivers and bays on these maps is worth making, apart from this question, because it throws considerable light on the cartography of the Gulf Coast.

As will be seen, the legend "Rio del Espíritu Santo" first appears in the northwestern corner of the Gulf, on the so-called Pineda map of 1520 or thereabouts.[1] The first printed map on which the course of this river is shown is that accompanying Cortés' second letter to Charles V. The bay into which it empties is unnamed, but afterwards, on the Salviati map of 1525-1527, it is called "Mar pequeña," small sea, because of its size. This legend is repeated on the Weimar map, the two maps of Ribero, and the Wolfenbüttel map. On the latter there is another name not found on the Weimar but repeated on nearly all the subsequent maps: "ostial." From Oviedo we learn that on the lost

---

[1] In this section bibliographical references are given to those maps which will not be mentioned later.

Chaves map of 1536, there was a descriptive legend not inscribed by previous mapmakers:

> Twenty leagues farther is the Cabo de la Cruz, near the mouth (en el embocamiento) of the Rio del Espíritu Santo, at latitude thirty-nine [*i.e.*, 29] degrees and two thirds of a degree on this side [scl. north] of the equator. . . . From the bay and headland (cabo) of the Puerto del Espíritu Santo to the Rio de Flores, there are seventy leagues [along the coast] running from east to west, and within the said bay is the Rio del Espíritu Santo; that bay [ensenada] is called Mar pequeña on the map. In the bay itself, from the Rio del Espíritu Santo to the *Culata* or end of the said bay or mar pequeña, east to west, there are twenty leagues, and the width is ten or twelve. Now then, from the mouth of this bay to the Rio de Flores towards the east, they count sixty leagues. . . .[2]

We do not know whether Chaves had the legend "Bahía del Puerto del Espíritu Santo" on his map, or whether these words are merely part of Oviedo's explanation; but six years later, on the world map of Alonso de Santa Cruz, the legend "Culata" is inscribed at the east end of the Mar pequeña, and the author of the so-called De Soto map, believed to have been Santa Cruz, also wrote inside the Mar pequeña "baya del espiritu Santo." This latter map is clearly posterior to the expedition of 1539-1543.

In the late fifties of the sixteenth century, one of the Agnese manuscript maps combined the two names thus: "B. del Spirito S^{to} delle Colata."[3] The Italian artist is here expressing cartographically a passage of López de Gómara's book on the conquest of Mexico, published at Saragossa in 1552, which reads as follows: "From the Rio de Flores there are seventy leagues to the Bahía del Espíritu Santo, which also goes by the name of *la Culata,* and which measures thirty leagues. From this bay which lies at latitude 29° there are more than seventy leagues to the

---

[2]G. F. de Oviedo y Valdés, *Historia General y Natural de las Indias, Islas y Tierra-firme del Mar Océano,* 4 volumes, Madrid, 1851-1855, II, 143.

[3]Reproduced in colors in K. Kretschmer, *Die Entdeckung Amerika's in ihrer Bedeutung für die Geschichte des Weltbildes,* Berlin, 1892, *Atlas,* Taf. XXV; black and white reproduction in G. P. Winship, "The Coronado Expedition, 1540-1542," published by the U. S. Bureau of American Ethnology, in the Fourteenth *Annual* Report, 1892-1893, Washington, D. C., 1896, pt. 1, between pp. 360-361.

Rio de Pescadores."[4] Evidently López de Gómara mistakenly thought that the name used by the Spanish geographers to designate the end of the Mar pequeña was the name of the bay itself. There is no mention of the Rio del Espíritu Santo. Though this name appears in the Latin translation of López de Gómara's book, it is clearly a mistranslation of the original Spanish.[5] Both Mercator and Ortelius repeated López de Gómara's mistake, but they inscribed "Rio del Espíritu Santo" on their maps.

Thus we see that thirty years after the appearance of the Rio del Espíritu Santo, the bay at its mouth, which was originally nameless had at first been called "Mar pequeña," then Bahía del Espíritu Santo, and finally appeared under both names.

On the west coast of the Floridan peninsula, the process is just the reverse. There we find first a Bahía del Espíritu Santo, without the corresponding river, and only later does a *rio* of that name appear.

It is well known that the bay in Florida where De Soto disembarked was called Bahía del Espíritu Santo because of the day on which land was sighted, Pentecost Sunday, May 25, 1539.[6] In 1554, a Portuguese mapmaker, Lopo Homem, drew at Lisbon a map of the world showing a "b. del espirito Sto" on the west coast of Florida.[7] As far as I know, this is the oldest extant

---

[4]"Del Rio de Flores ay setenta leguas a la Baia del Espiritu Santo, a quien llaman por otro nombre, la Culata, que boja treynta leguas. Desta baia, que esta en veyntinueue grados, ay mas de setenta leguas al rio de Pescadores," Francisco López de Gómara, *Primera y segunda parte de la historia general de las Indias,* . . . Saragossa, 1553², Fol. viii.

[5]"A flumine Florum ad sinum S. Spiritus. 70. miliaria numerãtur, quod Flumen alio nomine Culata dicitur, & 30 miliar. se extendit. Ab hoc sinu, qui est sub latitudine 29. graduum, ad fluuium Piscatorum usq; habentur vltra 70. miliaria;" *Cosmographia, siue Descriptio universi Orbis, Petri Apiani & Gemmae Frisij* . . . , Antwerp, 1584, 161. Cf. the sketch of Claude Delisle in Archives Nationales (AN), JJ, 75-137, and the note thereon by Joseph-Nicolas Delisle.

[6]*Relaçam verdadeira* . . . *feita per hũ fidalgo Deluas,* Evora, 1557, facsimile reproduction by J. A. Robertson, 2 volumes, Deland, Fla., 1932-1933, I, xix; Garcilaso de la Vega, *La Florida del Ynca. Historia del Adelantado Hernando de Soto,* . . . Lisbon, 1605, 28v.

[7][World map.] "Lopo Home[m] cosmografo caualeiro fidalgo del rei nosso Snor me fez e[m] lixboa Era de 1554 Annos," reproduced in G. Caraci, *Tabulae Geographicae Vetustiores in Italia Adservatae,* 3 volumes,

map bearing this legend at this point. There is conclusive evidence to show, however, that Homem followed a Spanish model. Instead of repeating Bahía del Espíritu Santo in the northwest corner of the Gulf Coast, he there employs the legends "R. del Espiritu Santo," "mar pequeno," and "culata." We know from copies of maps made on the Sevilian padrón, which found their way to France before the middle of the sixteenth century, that these maps were known in Portugal at a still earlier date. Even a superficial examination of the French copies shows that they were directly made not from Spanish models but from Portuguese versions thereof. The close affinity between Spanish and Portuguese rendered translation unnecessary, but the original gender of the Spanish adjectives is altered to make them agree with the Portuguese gender of the nouns, whenever these nouns have a different gender in the two languages. Thus, for instance, we find "mar pequeno" for *mar pequeña*. There are other linguistic changes: "praia" takes the place of *plaia*, "pracel" of *placel*; some draughtsmen consistently used "do," the contracted masculine Portuguese article where the Spanish has "del," but this was not so generally done as would appear from the practice of Homem. All of them, as a rule, wrote "lh" and "nh" for *ll* and *ñ*; and most followed the sixteenth-century Portuguese practice of spelling with an *m* words which were then written with an *n* in Spanish as they now are in modern Portuguese, v.g., *escomdido, homdo, samto,* etc. Finally we find the two liquids *l* and *r* transposed in such words as "froles" for *flores,* "Frolida" for *Florida.*

Since a Bahía del Espíritu Santo located on the west coast of Florida appears on a Portuguese map as early as 1554, and since it is clear that Portuguese mapmakers used—at least for this section of America—Spanish maps as models, we may conclude that Spanish mapmakers had inscribed a Bahía del Espíritu Santo on the west coast of Florida before the middle of the sixteenth century. Moreover, there is definite evidence that a decade later Spanish mapmakers began to inscribe two bahías del Espíritu Santo on their maps: one on the northwest corner of the Gulf

---

Florence, 1926-1932, I, Pl. I-IX, the Gulf of Mexico is Pl. II. On this cartographer, cf. A. Corteasão, *Cartografia e cartógrafos portugueses dos séculos XV e XVI,* 2 volumes, Lisbon, 1935, II, 7-28. This map of 1554 is thought by some to have been based on the lost Chaves map of 1536.

and another at or near De Soto's landing place on the west coast of Florida. Thus Francisco López de Velasco in his description of the "Gulf of New Spain . . . according to the maps of Santa Cruz," wrote: "The direction of the northern coast of this gulf runs for 250 leagues west from Florida, it then forms a southwestward arc to Pánuco, more than one hundred leagues long . . .";[8] and before describing the "Rio and Bahía del Espíritu Santo" on the north coast of the Gulf, he speaks of the west shore line of the Floridan peninsula as follows:

> Bahía del Espíritu Santo, the limit of the province (gobernación) of Florida, at latitude 29°, twenty or thirty leagues west of the Bahía Tocobaga. La bahía de Tocobaga also called [Bahía] del Espíritu Santo or de Miruelo, is at latitude 29°30'. . . . From Tocobaga to Tampa there are thirty leagues, the coast runs south-by-east. . . . At the entrance of the Bahía de Tampa—which may be the Bahía Honda—is a sand bank which is shown on old maps. The bay is large, with an opening which measures about three leagues, full of shoals covered by water. . . . From the above mentioned [Tampa] bay, there are twelve leagues to Carlos. . . . The Bahía de Carlos is called by the Indians Escampaba, after a cacique of this name . . . and seems to be the same as the bay called Juan Ponce, because he landed there in 1515 . . . it is situated at latitude 26°30'. . . .[9]

The date of "the maps of Santa Cruz" spoken of by López de Velasco can only be approximately ascertained. When he was appointed *cosmographo mayor* in 1572, he received from the government the maps, manuscripts, and papers of Santa Cruz, his predecessor in that office, who had died in 1567.[10] The latest map of Santa Cruz, which can be dated on other evidence, that of the Gulf of Mexico in the *Islario*, was made after 1560; on it are found the names "Bahía Honda" and "Bahía de Juan Ponce" on the west coast of Florida. The fact that López de Velasco found them replaced by the names "Bahía de Tampa" and "Bahía de Carlos" shows that the Santa Cruz maps to which he refers were made between 1560 and 1567.

The Florida Bahía del Espíritu Santo is found on a map of

---

[8] J. López de Velasco, *Geographia y Descripción Universal de las Indias*, Madrid, 1894, 180.

[9] *Id., ibid.*, 162-164.

[10] B. Boston, "The 'De Soto Map'," *Mid-America*, XXIII, 1941, 242-243.

Diogo Homem of 1568,[11] and on one of Oliva of 1580.[12] It appears for the first time in print on the Hierónymo de Chaves map published by Ortelius in his *Theatrum Orbis Terrarum,* 1584, and the legend is repeated on the copies of this map by Wytfliet, 1597, and Matal, 1602; also by Hondius on his world map of 1611, and by a few others. For some reason this Bahía del Espíritu Santo on the Florida coast did not obtain general recognition at the time; it is not found on many maps of the latter part of the sixteenth century, and only reappears at the very end of the seventeenth century. The name which is found, although it had a late start, is "R. or Rio del Espíritu Santo."

This latter name is first met with on an anonymous undated map of the Gulf, which was made after 1561.[13] It may be that the "R" on this map, which stands for Rio, is a mistake of the draughtsman for the initial letter of Bahía. The next time the legend appears on the west coast of Florida is on Blaeu's world map of 1605. From this date on, it is found with increasing frequency on printed maps of the first seventy-five years of the seventeenth century. Blaeu himself repeated it on all the maps which he himself or his son published, and since the French mapmakers following Nicolas Sanson's lead used Dutch maps as a basis for theirs, the Rio del Espíritu Santo on the west coast of Florida is found on French maps too. The following observations should be made, however. Sometimes the course of the river is shown, but no name is given to it; sometimes the name is given to what purports to be the mouth of the river, but the course is not shown; finally, on some maps, both name and course are shown.

On the so-called De Soto map there is, as noted above, a Bahía del Espíritu Santo in the northwest corner of the Gulf, and López de Gómara in his description of the Gulf also speaks of a Bahía del Espíritu Santo there. During the next seventy years, although

---

[11]Facsimile reproduction in V. Hantzsch and L. Schmidt, *Kartographische Denkmäler zur Entdeckungsgeschichte von Amerika, Asien, Australien und Afrika,* Leipzig, 1903, Taf. V.

[12]Photograph in the Karpinski Collection from the Bibliotheca Real, Madrid, 2-K-8.

[13]Carta de las Antillas, Seno Mejicano y Costas de Tierra Firme y de la América Setentrional in *Cartas de Indias,* Madrid, 1877.

the bay is always shown, the legend does not appear on any drawn or printed map, until the publication of the *Nieuwe Wereldt oft Beschrijvinghe van West-Indien* by De Laet in 1625. The reason for this omission is that the mapmakers of this period were merely reproducing manuscript or printed maps on which the bay was left unnamed, whereas De Laet has his maps drawn according to written accounts which he summarized, and it is clear that to draw the Gulf Coast a modified López de Gómara text was used by the draughtsman.

Although all the maps published by the Blaeu firm during the first half of the seventeenth century have two rios del Espíritu Santo, one on the west coast of Florida and the other at the northwest end of the Gulf, yet on only one map,[14] published by John Blaeu, is the bay into which the latter discharges itself labeled *Baya de Sp. Santo,* and even this map is one of those which had been previously published in Hondius' *Nouveau Theâtre du Monde* of 1639.

On Nicolas Sanson's 1650 map of North America, the nomenclature of the Gulf Coast from the Florida Keys to the Bahía del Espíritu Santo is lifted bodily from De Laet's map entitled *Florida et Regiones Vicinae*; on this map, and therefore also on Sanson's, a "R. de Spiritu Santo" disembogues into a bay on the west coast of Florida, but no name is given to any of the six rivers emptying into the *B. de Spiritu S*^{to} in the northwest corner of the Gulf. Six years later Sanson issued another map, the characteristic trait of which consists in his having superimposed the hydrography and the nomenclature of the western part of the Chaves map published by Ortelius in 1584, on the hydrography and the nomenclature of the De Laet map of 1625. His 1656 map therefore, contains a Rio del Espíritu Santo emptying into a Bahía del Espíritu Santo in the northwest corner of the Gulf, and another Rio del Espíritu Santo the mouth of which is on the west coast of Florida.

As a result of this superimposing, Sanson's 1656 map is evidently worthless in many respects, yet so great was his popularity as a mapmaker that after the discovery of the Mississippi some draughtsmen traced the course of the newly found river on one

---

[14]*America Septentrionalis,* Amsterdam, 1641. Cf. Phillips' note in Lowery, *A Descriptive List,* 139.

of Sanson's North American maps, while others more judicious, made use of Dutch maps instead.

In 1674, the year following the discovery of the Mississippi, but before any news of it could have reached Paris, Hubert Jaillot published a map of North America quite different from those of Sanson. Into the Bahía del Espíritu Santo in the northwest corner of the Gulf a R. de Espíritu Santo discharges itself, but this river is insignificant in comparison with the Chucagua which empties into the same bay. Five years later, in 1679, Duval's new conception of the geography of the Gulf Coast appeared in print. He has two "Rivières du St. Esprit" emptying into the northwestern bay, one coming from the northwest, and the other from the northeast.

From the above brief review of maps by geographers of every nationality in western Europe, it seems clear that they either dutifully copied old maps or indiscriminately inserted nomenclature and geographical features found on several maps into their own. Some have one Rio del Espíritu Santo, others have two; some legend the western one, others the eastern one, and still others write out the name along the course of both.

The news of La Salle's descent of the Mississippi to the sea reached Paris at the time when Coronelli was making the gores of his famous globe.[15] On his map of 1688—a variant of the gores of 1683—the Mississippi empties into the Gulf five degrees west of the "Bahia de lo Espu Santo, ò Mar pequeno," and into this bay flow the waters of a "Rio de lo Spiritu Santo," only the mouth of which is shown, while a 400-mile-long "R. del Spo Santo" discharges itself into Apalachee Bay. These peculiar features were repeated by Franquelin in 1684, when he redrew La Salle's map of the Mississippi.

It would be pointless to extend this review further or to enter into greater detail, since many of these maps will be examined in subsequent sections of this essay in connection with the sources used by the mapmakers. It will thus become clear that there is hardly any valid evidence for asserting that the Rio del Espíritu Santo of the Spanish geographers is identical with the Mississippi.

---

[15] Cf. J. Delanglez, *Some La Salle Journeys*, Chicago, 1938, 37-39; id., *Hennepin's Description of Louisiana*, Chicago, 1941, 113-114; id., "La Salle's Expedition of 1682," *Mid-America*, XXII, 1940, 22-23.

We shall only remark here that on his early maps and globes published under the name of his son Guillaume, Claude Delisle also has a Baye du St. Esprit in the northwestern corner of the Gulf,[16] but beginning with the map of 1703,[17] all traces of such a bay disappear, and the only Baye du St. Esprit which he mentions is now located on the west coast of the Floridan peninsula where Nicolas De Fer had placed it in 1699.[18]

The hundreds of maps of the Gulf Coast and of the country north of it which were examined in preparing this essay have been divided, for the sake of convenience, into several groups. We have grouped together, for instance, those showing the coast line and merely indicating the mouth of the rivers; another group shows the course of some rivers, but this delineation is clearly the result of the mapmaker's fancy, and not of actual exploration; a third group of maps has the interior in greater detail, but in their case it is quite difficult to separate fancy from fact, or to ascertain the source of the draughtsman's information. In these and in the other groups of maps, we have attempted to isolate the essential geographical features and the fundamental nomenclature of the matrix map which served as a model. We have been careful to note additions occurring on some maps which in many cases have no other foundation than the draughtsman's imagination, or omissions which may be attributed to the inferior skill of the copyist or to exigencies of space. These changes obviously make no real difference. The additions do not argue progress in geographical knowledge any more than the omissions are a proof of priority.

An important remark must be made at this point. When in the course of the following discussion two or more maps are said to be the same, it is understood that the similarity only applies to definite geographical features in the restricted area north of the Gulf Coast line from the Florida Keys to Tampico.

---

[16] See the "Carte de la Nouvelle France et des Pays voisins, 1696," in AN, JJ, 75-130, as well as the two variants, AN, JJ, 75-128; *Globe Terrestre . . . ,* Paris, 1700; *L'Amerique Septentrionale,* Paris, 1700.

[17] *Carte du Mexique et de la Floride . . . du Cours et des Environs de la Riviere de Mississippi,* Paris, 1703.

[18] "Carte de la Coste et des Environs du Fleuve de Mississipi, 1699," in Bibliothèque du Service Hydrographique (SHB), C 4040-2; this map is not signed, but compare it with De Fer's of 1701 in SHB, C 4040-5.

Besides these maps which, as we have said before, are one of the best means of learning what was known of the geography of the Lower Mississippi Valley and the adjacent territory before 1673, written descriptions of the country have also been made use of in order to ascertain whether the Rio del Espíritu Santo of the Spanish geographers was the Mississippi. The time limit of our enquiry, moreover, has been extended beyond 1673, because after this date the French and the Spaniards were still in doubt about the identity of the Mississippi, although the French had explored its lower length to the Gulf in 1682, and the Spaniards rediscovered its mouth from the sea a few years later.

After these preliminary remarks which contain an outline of the plan followed in this essay, we shall now examine the first group of maps, namely, those which show only the coast line and the mouth of the rivers emptying into the Gulf.

II

THE FIRST GROUP OF MAPS—THE PADRÓN

The maps of Cantino,[1] Canerio,[2] Waldseemüller,[3] and the Admíral's map[4] are quite useless for the present discussion. It is by no means certain that the continent shown to the northwest of "Isabelle Insula" is the present-day territory which roughly includes Mississippi, Alabama, Georgia, and South Carolina, or that the island northwest of "Spagnola" on the Ruysch map[5] represents the same region; and there is little likelihood that the peninsula jutting out into the ocean toward the southeast on the first four of these maps represents Florida. If the gulf shown on them is indeed the Gulf of Mexico, and if these maps are cartographical expressions of actual explorations, it still remains to be explained why the mapmakers studded this gulf with so many islands. Dawson's theory,[6] which ingeniously harmonizes

---

[1]Full-size photographic reproduction in E. L. Stevenson, *Maps Illustrating Early Discovery and Exploration in America (1502-1530)*, New Brunswick, N. J., 1906, no. 1. The American part is reproduced in colors in H. Harrisse, *Les Corte-Real et leurs voyages au Nouveau-Monde,* Paris, 1883; also in black and white, greatly reduced, in Harrisse, *The Discovery of North America,* Paris, 1892, 79, etc. A good bibliography on this and other early maps will be found in Lowery's *Descriptive List of Maps of the Spanish Possessions.*—The map derived its name from an Italian inscription on the back: "Carta da nauigar per le Isole nouam^{te} tr[ouate] in le parte de l'India: dono Alberto Cantino al S. Duca Hercole." Cantino was an ambassador from Hercules d'Este, duke of Ferrara, to the King of Portugal.

[2]E. L. Stevenson, *Marine World Chart of Nicolo Canerio Januensis. 1502 (circa). A Critical Study with Facsimile,* New York, 1908. The northern part of the Western Hemisphere in Harrisse, *Discovery of North America,* 306; North and South America in Kretschmer, *Atlas,* Taf. VIII.

[3]J. Fischer and R. von Wieser, eds., *The Oldest Map with the name America of the year 1507 and the Carta Marina of the year 1516 by M. Waldseemüller (Ilacomilus),* Innsbruck, 1903.

[4]"Tabula Terre Nove," in Ptolemy, *Geographiae opus,* Augsburg, 1513. The map has often been reproduced, v.g., in Winsor, *Narrative and Critical History of America,* 8 volumes, Boston and New York, °1884-1889, II, 12, Kretschmer, Taf. XII, no. 1, etc.

[5]"Vniversalior Cogniti Orbis Tabula ex recentibus confecta observationibvs," in *Geographiae Cl. Ptolemaei,* Rome, 1508, reproduced in Winsor, *Narrative and Critical History,* III, 9.

[6]S. E. Dawson, *The Saint Lawrence Basin and its Border-Lands,* London, 1905, 54.

these maps with documentary evidence, rightly insists that the hypothesis of "unknown navigators" is too easy a solution of cartographical puzzles.

The earliest map to show the contours of the Gulf is a very crude sketch, known as the Pineda map,[7] which, it is believed was sent to Spain by Francisco de Garay.[8] This sketch is a mere outline on which neither latitudes nor longitudes are marked, but for all its crudeness, considering that it is very probably the first drawing made from actual observation, it gives a good delineation of the coast of the whole Gulf from the Florida Keys to Yucatan. The mouths of three unnamed rivers are indicated between "Rio Panuco" and "Rio d. Espu. Santo." At the mouth of the latter river, the mapmaker began to draw a large bay, to which, however, he gave no name.

With regard to a possible identification of the Rio del Espíritu Santo with the Mississippi on the basis of this sketch, the following should be said: We have no account by Pineda of his journey along the coast; we do not even have the petition sent by Garay to the court; all we have is the royal *cédula* of 1521 authorizing the governor of Jamaica to colonize the country. In this *cédula* there is a vague reference to a "drawing" made by someone unknown. It has been gratuitously surmised that this "drawing" is the Pineda sketch, on which the mouth of a river is legended

---

[7] Archivo General de Indias (AGI), Patronato, 1-1-1/26-16, photograph in the Karpinski Series of Reproductions. A facsimile of a re-drawing of the map in M. Fernández de Navarrete, *Colección de los Viajes y Descubrimientos que hicieron por mar los Españoles del Siglo XV . . .* , 5 volumes, Madrid, 1858-1880², III, facing p. 148; in N.A.E., Nordenskiöld, *Periplus, an Essay on the Early History of Charts and Sailing-directions,* Stockholm, 1897, 179; in Kretschmer, Taf. XIV, no. 6, etc.—*Infra,* pl. 1.

[8] For the antecedents of the voyage of "Alonso Alvares de Pineda o Pinedo," as Díaz del Castillo calls him, cf. *Historia verdadera de la Conquista de la Nueva España,* 3 volumes, Mexico City, 1939, I, 210, 211. For his route, cf. the cédula of 1521, in Fernández de Navarrete, *Colección de los Viajes,* III, 148-149, and Garay's declaration of 1523, in *Colección de documentos inéditos relativos al descubrimiento, conquisto y colonización de las posesiones Españoles en América y Oceania,* 42 volumes, Madrid, 1864-1884, XXVIII, 500. Cf. the interpretation of Harrisse, *Discovery of North America,* 167 ff., and that of D. L. Molinari, *El Nacimiento del Nuevo Mondo, 1492-1534, Historia y Cartografía,* Buenos Aires, ᶜ1941.

"Rio del Espíritu Santo." The river mentioned, but not named, in the royal document of 1521 is described as follows:

> They entered a very large and mighty [literally, carrying much water] river. At its entrance, you say, they found a large town where they remained forty days careening their vessels. The people of the land were very friendly toward the Spaniards who went with the fleet, trading with them of what they had. They sailed up the river a distance of six leagues and found forty towns on either side. They admired very much all of the country along which they coasted and which they discovered, the land, the harbors, and the rivers, as appears from a drawing (*como por una figura . . . parecia*), which has been delivered unto Us on your behalf by the pilots who went with the said fleet. Thereby it appears that the said *adelantados* Diego Velasquez and Juan Ponce de León and yourself, have discovered the mainland and the coast thereof, and that the coast and mainland which you discovered was given the name Provincia de Amichel. . . .[9]

If one does not try to read into the text what is not there, all that we learn from this description is that a large river empties into the Gulf. No mention is made of its passing through a bay as indicated on the Pineda map, and hence there is no reason for asserting that this river is the Rio del Espíritu Santo as shown on this map. Besides, if one considers the position of the Rio del Espíritu Santo on the map, it can hardly be the Mississippi. For, although neither coordinates nor scale are given, the mouth of the river is located near the southward curve of the coast line, much farther west than the actual position of the delta, which, as we know, divides the northern coast of the Gulf into two nearly equal halves. Furthermore, it is difficult to see how the description of the river in the *cédula* could apply to the mouth of the Mississippi at any historical period;[10] it is also difficult to see how there could have been a large town at the entrance of the river, and especially how there could have been forty towns on the banks of the Mississippi over a distance of six Spanish sea leagues from its mouth, if we take into consideration the geo-

---

[9] Royal *cédula* of 1521, in Fernandez de Navarrete, *Collección de los Viajes*, III, 149; cf. Garay's declaration in *Collección de documentos inéditos*, XXVIII, 500.

[10] Cf. the reasoning of Scaife, *op. cit.,* 145 ff.

logical formation of the delta. It is generally admitted that during the past four centuries the delta has advanced ten miles into the Gulf. Hence the forty towns would have been clustered along the banks over a distance of twenty-five miles, as far as what we now call "The Jump." Anyone who has seen the Mississippi twenty-five miles above the passes will hesitate before identifying it with the river described in the *cédula* which asserts that there were twenty towns, a mile apart, on both banks of the river.

We shall next consider what may be called the Cortés map, although it properly belongs to the second group. In 1524, the first Latin edition of Cortés second letter of 1520 was published at Nuremberg.[11] This edition contains a plan of Mexico City and also a map of the Gulf. On it, between the Rio Pánuco and the Rio del Espíritu Santo, five names are inscribed, three of which are descriptive rather than geographical: "Tamacho pruicia," "R la palma," "R de Arboledas," "p° de Arrecifos," and "Rio del spiritu sancto." Many difficulties which arise in connection with the Pineda drawing could be solved if we knew the origin of the manuscript sketch used for this map by the German engraver. Cortés wrote in the letter: "I also prayed Montezuma to tell me if on the seacoast there was any river or bay where ships could enter safely, and he answered me that he did not know, but that he would have the coast drawn for me, with its bays and rivers, and that I might send the Spaniards to see them and that he would give me people to guide and take them;

---

[11]*Praeclara Ferdinãndi Cortesii de Noua maris Oceani Hyspanica Narratio,* Nuremberg, 1524. Six months later, there appeared an Italian translation of this letter, *La preclara Narratione di Ferdinando Cortese della Nuoua Hispagna del Mare Oceano,* Venice, 1524. The Venetian engraver traced the outline of the map, and inserted the place-names—some of which he translated into Italian—in Latin characters. He was evidently unfamiliar with Gothic characters; he transcribed, for instance, "Rio de dos bocas," by "Rio de dos Gotas." The German engraver began to write the legend "Rio panüco" too close to the coast line, so that the last syllable "co" had to be inscribed between the two branches of the river. The legend "laoton" meant to designate the region west of the river is so close to the syllable "co" as to make it appear one word, and in fact the Italian engraver wrote "Colaoton." This mistaken reading is found on many subsequent maps, either because later draughtsmen erred in the same manner as the Venetian or because they copied the latter's map.—*Infra*, pl. 3. (II)

and thus he did."[12] It is doubtful, however, if this drawing showed the coast line from the Rio Pánuco to the Rio del Espíritu Santo, and it is also doubtful if the Spaniards explored the coast line from their base at Vera Cruz to the Punto de Arrecifes,[13] which latter is still about fifty Spanish sea leagues from the Rio del Espíritu Santo on the map.

The sketch used by the German engraver was obviously similar to the Pineda drawing, and the additional nomenclature on the printed map, although somewhat vague, seems to be genuine. It is known that some of Pineda's men were captured by Cortés, and these may have supplied the conquistador with the information necessary to draw the coast line as well as with the additional names which are found on the printed map. Cortés would naturally omit the two legends in which the exploration of the whole coast line from Apalachee Bay to the Gulf of Honduras is credited to Garay and Diego Velásquez. With regard to the position of the Rio del Espíritu Santo, we may note that on the Cortés map its mouth is as much to the east of the real position of the Mississippi as it is to the west on the Pineda sketch.

The nomenclature of the Gulf Coast from Florida to the Rio Pánuco on the Visconte de Maggiollo map of 1527[14] is similar to that on the Cortés map. The "Rio De Spiritu Stõ" is placed where the coast begins to slope toward the southwest, and there is no sharp bend on the coast line until the Rio Pánuco, the name of which is not inscribed, but its mouth is represented by a huge bay twice as long as that into which the Rio del Espíritu Santo empties. The legend "lacoto" between the two rivers which join to form the Rio Pánuco estuary is evidently a misreading of the word "laoton" on the Cortés map.

The Maggiollo map is mentioned here not for any intrinsic merits, but because it has certain points of resemblance with the Verrazano map.[15] On both these maps the Yucatan peninsula is

---

[12] F. A. MacNutt, *Fernando Cortés his five letters of Relation to the Emperor Charles V,* 2 volumes, Cleveland, 1908, I, 244-245.

[13] Cf. MacNutt, *ibid.,* 245, and Barcia, *Ensayo cronologico,* Madrid, 1723, 7.

[14] [Map of the World], in Stevenson, *Maps Illustrating,* no. 10; Western Hemisphere in Harrisse, *Discovery of North America,* between pp. 216-217; North and Central America, in Kretschmer, Taf. XIV, no. 7.

[15] It bears the inscription "Hyeronimus de Verraxano faciebat," Steven-

represented as an island as it is on the Cortés map, and on both maps Florida is joined to a northern continent by a narrow isthmus. The latitudes marked on both maps are hopelessly inaccurate. Thus, the southernmost tip of the Florida peninsula on the Maggiollo map is at latitude 20°, five degrees too low, and the Tropic of Cancer runs just north of the bay into which the Rio del Espíritu Santo empties; on the Verrazano map, the southernmost end of Florida is at latitude 33°30', eight and a half degrees too high, and the northern coast of the Gulf runs along the 39th degree of latitude, *i.e.,* the parallel passing through the mouth of the Missouri River and Washington, D. C.

The nomenclature along the Gulf Coast on the Verrazano map is unique. Beginning at the western end of a large bay into which the Rio del Espíritu Santo empties, and reading eastward, we have the following names which are found on no other maps: "Onari," "tabarca," "Golfo dasbasa," the last presumably the name of the bay itself; the mouth of a river is labeled "C. formiter," and next come the two words "quippia" and "Corominel." After this begins an inscription which the mapmaker saw on the Pineda map or on some copy of it. The Pineda map as we have it today has: "desde aquj començo || a descubrir fran^{co} de || garay." The Verrazano variant of this is as follows: "qui comincio || a descoprit || franc° || degarra." And immediately after this inscription we have the following words: "ultimo || della || nova || Hispania."

The Maggiollo map and especially the Verrazano are of no value for ascertaining the state of geographical knowledge of the Gulf Coast between 1527 and 1529, and still less for ascertaining whether the Rio del Espíritu Santo of the Spaniards is the Mississippi. On the Verrazano map the Rio del Espíritu Santo is not even mentioned.

Before coming to the maps based on the padrón made by the official Spanish cartographers, mention should be made of a chart known as the Salviati map,[16] so called because the coat of arms of this distinguished Italian family is found twice on the map.

---

son, *Maps Illustrating,* no. 12; a good reproduction of North America is in C. O. Paullin, *Atlas of the Historical Geography of the United States,* New York, 1932, Pl. XIII.

[16]Stevenson, *Maps Illustrating,* no. 7.

Giovanni Salviati was papal nuncio in Spain from 1525 to 1530, and it is surmised that the map was made during that time. The present writer believes that it was based on a prototype anterior to the padrón of 1527-1529. This opinion was arrived at by comparing the nomenclature from the Rio Pánuco to the Rio del Espíritu Santo on this Salviati map with that of the maps based on the Seville padrón of about 1527. The nomenclature for this section of the Gulf Coast is similar in both, and it may well express cartographically the reports of the pilots who sailed along the coast between 1520 and 1525. The fuller nomenclature on the maps of 1527 and 1529 is accounted for by the fact that the Spanish geographers received further pilots' reports from 1526 to 1529. That these new names were not invented by the geographers in Seville, but that they represent the results of actual observations, is practically certain. These men had no intention to deceive; it was of paramount importance that they should have as accurate knowledge as possible of the coast line of the New World. Nor is it reasonable to think that the pilots filled up their official sailing charts at random with names of places which they had not seen. This would have been too dangerous, for when these same pilots returned to those shores they would be jeopardizing their own lives as well as the lives of their crew by inaccurate records.

This conclusion is derived from what we know of the workings of the geographical department of the Casa de Contratación in Seville. A brief account of the origin and mechanism of this organization will make clear the grounds for the deductions of the preceding paragraph. It will also explain the similarity of so many maps of the Gulf Coast and will help to a solution of the main question at issue, namely, whether the Rio del Espíritu Santo of the Spanish geographers is the Mississippi. Once we know the genesis of the padrón, *i.e.*, of the basic map or prototype, and once we know whether it is trustworthy and accurate, it only remains to show that this matrix was copied again and again by subsequent mapmakers who introduced occasional changes in the nomenclature. As will be seen, these changes were accidental and do not affect the position of the main geographical features of the coast line, among them the position of the Rio del Espíritu Santo.

Before Columbus left for his second voyage, the Council of

the Indies had been established as a special department of the government, with direct control over the affairs of the New World. Ten years later, in 1503, an agency independent of the Council was founded at Seville, the Casa de Contratación, which had jurisdiction over the trade between the mother country and the Spanish dominions. New discoveries and conquests gave rise to new administrative problems; and in order to handle them more efficiently, the Casa was divided into departments. The geographical department, established in August 1508, was composed of pilots, cartographers, professors of cosmography or navigation, and technicians who designed maps and sailing charts, and was headed by a chairman who had the title of *piloto mayor*.[17]

In the decree creating this hydrographic service, the king declared that upon being informed that "there were many models of charts by various masters . . . which greatly differed one from the other,"[18] he therefore ordered a patent map to be drawn which was to be called *Padrón Real*. One of the reasons for having this pattern made was the utility of accurate maps for navigation and trade, since good maps of the coast line of the fast expanding Spanish empire would notably lessen the number of shipwrecks.

This order of Ferdinand was carried out by a commission composed of the ablest members of the department presided over by the *piloto mayor*. The model which they were to make was to include all the land and isles of the Indies then discovered which belonged to the Spanish sovereign. This general map was to be official, and all pilots were forbidden to use any other under a

---

[17] More detailed accounts concerning the Casa de Contratación will be found in M. Fernández de Navarrete, *Disertación sobre la Historia de Náutica*, Madrid, 1816, 132 ff.; Harrisse, *Discovery of North America*, 13 ff., and 256 ff.; G. Latorre, "La Cartografía Colonial Americana," in *Boletín del Centro de Estudios Americanistas*, año III (1915), no. 6, 1-10; nos. 9 and 10, 1-14; E. L. Stevenson, "The Geographical Activities of the Casa de la Contratación," in *Annals* of the Association of American Geographers, XVII, 1927, 39-59; etc. All these studies are based on Joseph de Veitia Linaje, *Norte de la Contratación de las Indias Occidentales*, 2 volumes in one, Seville, 1672, 139 ff.; the various ordinances and royal *cédulas* will be found in *Recopilación de Leyes de Reynos de las Indias*, 4 volumes, Madrid, 1756², III, lib. IX, tit. xxiii, 285 ff.

[18] *Cédula* of August 6, 1508, addressed to Amerigo Vespucci, in Fernández de Navarrete, *Colección de los Viajes*, III, 303.

penalty of fifty doubloons. They were also enjoined to mark on the copy used during their voyages all lands, isles, bays, harbors, and other things worthy of note, and they were commanded to communicate the chart so amended and annotated to the *piloto mayor* upon their return to Spain. The latter would then discuss with his colleagues of the Casa whether the information received would be inserted in the *Padrón Real*. The *piloto mayor* and certain crown pilots were allowed to make copies of the padrón or to have them made by the mapmakers of the department, and were allowed to sell them for their own benefit, according to a tariff determined by the Casa.[19]

The elements of the first model were borrowed from maps then current in Spain. But an ordinance of Charles V issued sixteen years later shows that the cosmographers of the Casa were remiss in carrying out the provisions of the decree of 1508. Either because the pilots failed to return their annotated maps, or because the members of the department neglected to insert in the padrón the geographical knowledge acquired as further discoveries were made, the official map, which should have embodied the best and latest information was considered inferior, in 1512, to that of Andrés de Morales.[20] At the time of Charles V's ordinance, just as before the decree of 1508, charts at variance with each other were still in circulation. They were being made and sold by pilots who were unwilling to submit their charts for approval to the *piloto mayor,* since they were encroaching upon his and other crown pilots' monopoly. The accuracy of these bootlegged maps depended on the fancy of their makers: latitudes varied, scales were changed, directions were altered, and coast delineations modified, all of which rendered navigation insecure.

This state of affairs lasted until 1526. On October 6 of that year, Charles V ordered Diego Ribero and the other cosmographers and pilots of the Casa to draw an official map, known henceforth as the *Padrón General*.[21] This official map was to be kept in the Casa. At the beginning of each year the *piloto mayor,* after

---

[19]Cf. the text taken from the Muñoz manuscripts in Harrisse, *Discovery of North America,* 15, note 8.

[20]Fernández de Navarrete, *Disertación sobre la Historia de la Náutica,* 138.

[21]*Colección de documentos inéditos,* XXXII, 512.

conferring with the cosmographers of the department and other persons versed in navigation, was to insert in it what had been learned during the course of the preceding year. Charles V also "authorized all professional cartographers residing at Seville to design and sell maps of the New World, with no other restriction than to have the same approved by the *piloto mayor* and the cosmographers of the Casa. He even permitted the *piloto mayor* himself to sell not only copies of the *Padrón General* but also maps and globes of his own making, provided that trade in such articles was not carried on within the city of Seville."[22] All maps were to be modeled after the *Padrón General* and no other. Members of the geographical department of the Casa transgressing this last provision were to be fined 50,000 *maravédis*, and suspended from their office at the king's pleasure.[23]

The *Padrón General* was not a completely new map. Since the *Padrón Real* had been compiled from the maps available in Spain, the geographers of the Casa used the *Padrón Real* as a base, complementing, correcting, amending, and bringing it up to date. We possess no copy of either of these two standard maps, the *Padrón Real* and the *Padrón General*, but a map of the world drawn in 1527 very likely represents the *Padrón General* as it appeared in that year. This map, known as the Weimar map or globe after the town where it is preserved,[24] is unfinished and only shows the east coast of the New World from Labrador to the Strait of Magellan. There are three other extant maps which were made within a few years after the decree of 1526, and which represent various states of the *Padrón General*. One, un-

---

[22]Harrisse, *Discovery of North America*, 16-17.
[23]*Recopilación de Leyes,* III, lib. IX, tit. xxiii, ley xii, 287v.
[24]Carta Vniversal en qve se contiene todo lo qve del Mvndo sea Descvb[ierto] fasta aora hizola vn cosmographo de sv magestad Anno M. D. XX. VII. en Sevilla. Reproduced in Stevenson, *Maps Illustrating*, no. 9. "It seems probable," says Stevenson, "that this map was constructed for Charles V, who took it to Rome, whence it was brought to Nuremberg in 1530 or thereabouts." It was acquired by the Grand Duke Karl August in 1811, and placed in the Grand Ducal Library. J. G. Kohl, who made a detailed study of this map as well as of that of 1529 by Ribero, *infra*, holds them to be two distinct works based on one *padrón, Die beiden ältesten General-Karten von Amerika,* Weimar, 1860. I am inclined to believe that Ribero is the author of both.—*Infra,* pl. 2. (I)

dated, unsigned, and unfinished, was discovered in Wolfenbüttel Castle, Germany;[25] the other two are signed by Diego Ribero, the cosmographer whom Charles V ordered to make the *Padrón General*. These two maps—or rather these two states of one map—were made in 1529.[26] One is in the Grand Ducal Library of Weimar with the 1527 map, while the other, sometimes referred to as the "Second Borgia Map" after the name of the donor, is in the library of the De Propaganda Fide college at Rome.[27]

On May 20, 1535, Isabella of Portugal ordered Fernando Columbus to assemble pilots and cosmographers in Seville, and to have the all important map which her husband, Charles V, had commanded Diego Ribero to make in 1526, executed at once.[28] According to Harrisse, this order was issued by the queen because Charles V's command "remained a dead letter for nine years,"[29] and because there was still no padrón in 1535. This interpretation is possible on the wording of the *cédula*, but there are good reasons for rejecting it. The Weimar map was made in Seville, as is specified in the title by "a cosmographer of His Majesty," who was very probably Ribero himself; and there are two other maps signed by Ribero—who had been commanded by Charles V to execute the *Padrón General*—which were also made in Seville in 1529. Moreover, all these maps are so similar in delineation and nomenclature, and the positions of towns and rivers are so nearly alike that one may legitimately conclude that they were all

---

[25] Photographic reproduction, actual size, in Stevenson, *Maps Illustrating*, no. 8.

[26] Carta Universal en que Se contiene todo lo que del mundo se ha descubierto fasta agora. Hizola Diego Ribero Cosmographo de Su Magestad: Año de 1529. This map, or part of it, has often been reproduced; the most satisfactory is the actual size photograph in Stevenson, *Maps Illustrating*, no. 11. Cf. J. G. Kohl, *Die beiden ältesten General-Karten,* and M. C. Sprengel, *Ueber Diego Riberos Welt-Karte von 1529,* Weimar, 1795.

[27] Carta Vniversal en que se contiene todo lo que del mundo se ha descubierto fasta agora. hizola Diego Ribero cosmographo de Su Magestad: Año de: 1529 e Seujlla. Reduced facsimile in colors published by W. Griggs, London, [1887]; on this reproduction cf. E.-T. Hamy, *Note sur la Mappemonde de Diego Ribero (1529) Conservée au Musée de la Propagande de Rome,* Paris, 1887.

[28] *Colección de documentos inéditos,* XXXII, 512-513.

[29] *Discovery of North America,* 16.

made on various states of a basic model which can only have been the *Padrón General*. Finally, it is hardly credible that the geographers of the Casa waited nine years before carrying out the order of the king.

With regard to these various official maps, the following may be said. Between 1509 and 1525, there was in Seville a matrix map called the *Padrón Real* compiled from maps available in Spain. At some time during these years the Gulf of Mexico was inserted in it. Whether the Turin, the Salviati, and other maps were based on it is a matter of conjecture. In 1527, the geographers of the Casa revised this chart, which was known thereafter as the *Padrón General*, and which is very probably the model followed by the author of the Weimar globe of 1527, and by Ribero for his maps of 1529. We know that in 1536, the year after Isabella of Portugal wrote to Fernando Columbus, a map was made by Alonso de Chaves which was almost certainly a copy of the *Padrón General* as it then existed in the Casa de Contratación at Seville. Hence although most of the magnificent cartographical output of the early Spanish geographers is lost, their extant charts enable us to form an idea of the appearance of the model kept in Seville, and thus we can ascertain what the geographers of the Casa de Contratación knew about the Gulf of Mexico, and about the rivers along the coast line which were sighted, named, and navigated by Spanish pilots and explorers.

A study of the accompanying table of the nomenclature of the Gulf Coast on four early Spanish maps will show that the Weimar map represents an earlier state of the *Padrón General*. If, as I believe, Ribero is the author of the Weimar map, it is likely that this earlier map is the *Padrón Real* with emendations. When, however, two years later, this able geographer drew his charts of 1529, the model map had undergone further changes. It is clear that the Wolfenbüttel map contains elements of the other two, plus a few legends from maps no longer extant. On the Weimar map, the Rio del Espíritu Santo is located on the northern coast line, two thirds of the distance between the Floridan peninsula and the point where the coast begins to bend southwestwards, 150 leagues from the east shore of Apalachee Bay. These leagues are Spanish sea leagues, numbering seventeen and a half to a degree of latitude; hence the river on this map is located near present-

NOMENCLATURE ON FOUR EARLY SPANISH MAPS OF THE GULF OF MEXICO

Weimar	Wolfenbüttel	Ribero - Weimar	Ribero - Borgia
Las palmas	R. de palmas costa tesa	costa tesa Rio de palmas	R. de Palmas
la madalena	C. bravo malabrigo	C. bravo malabrigo	C. bravo
R. scondido	R. escondido anegados playa baxa	R. escondido anegados playa baxa	R. escondido anegados p. baxa
R. de loro	R. de loro Tra de gigantes	R. del oro Tĩa de gigãtes	R. del oro tẽra de gigãtes
pº de arrecifes	C. sirto C. de cruz	C. sirto	C. sirto
C de ✠			
R. del Spiritu Sancto	R. del Espũ Sto ostial	R. del Espũ Stõ ostial	R. del espũ Stõ ostial
mar pequeña	mar pequena	mar pequeña	mar peqña
Matas // de S. Salvador	Motas de Salvador	Motas del Salvador	Motas del Salvador
Ancones	Ancones	Ancones	Ancones
Medanos		Medanos	
Aldea		aldea	
p. llana			
el canaveral	el canaveral	recifes	
R. de flores	R. de flores	R. de flores	R. de flores
Reciffes	Angla	recifes	Medanos
	praceles Ancones	poeles	p. de baxos
R. de nieves	R. de nieves	R. de nieves	R. de nieves
dende aqui descobrio frco de Garay		terra de Garay	dende aq. descubrio Garay canaveral c. de farallones
costa tesa	costa tessa	costa tessa	
R. de S. Juhan	R. de S. Juhan		R. de S. Juan
las matas			
ancon baxo	pº baxe navidade boque C. escondido R. de la concepcion		
	motas atalaya	R. de cõabicion motas atalaya	R. de cõabicion motas atalaya
b. de Juan ponce	b. de Juhan ponce	b. de juã ponce	b. de Juã ponce
R. de canoas		R. de canoas	R. esta para [?] R. de la paz
[R] de la paz			
R. de spana	playa	aguada	
	canico	canico	

day Galveston Bay. On the other three maps, which as noted above, represent either the amended *Padrón Real* or else an improved *Padrón General,* the bay into which the river disembogues is located at the same place, and the shape of the bay on the Ribero map of 1529 so closely resembles Galveston Bay that it would seem that the cosmographer drew it on reports of pilots who saw it.

Much information about the *Padrón General* as it existed in 1536 is found in Oviedo. This chronicler was greatly interested in maps. He seems to have known Ribero personally, and he was the friend of two other cosmographers of the Casa, Alonso de Chaves and Alonso de Santa Cruz. Whatever may be said about Oviedo's "prejudices," he had a more critical mind than many of his detractors, both contemporary and modern. In his *Historia General y Natural de las Indias,* he describes the American coast line from Labrador to the Strait of Magellan, according to the "latest maps by the cosmographer Alonso de Chaves in the year 1536, in conformity with the order of our sovereign, the Emperor, commanding learned and experienced persons, chosen for this purpose, to examine the *padrones* and the sailing charts."[30] He warns the reader that this map of Chaves does not represent the coast line north of latitude 51°30′, and that he used the Ribero map for describing the coast above this latitude. He makes many comparisons between the map of Ribero and that of Chaves, and corrects some of the data given by the latter with the help of the memoirs and maps of Santa Cruz. With regard to the coast of Argentine, for instance, about which the maps or Santa Cruz and Chaves are at variance, he accepts the data of the former in preference to those of the latter, because Santa Cruz actually saw this coast, having gone thither with Sebastian Cabot in 1526. With regard to the coast of Castilla del Oro, he discarded the distances given by Chaves, because "I spent the best part of my life" in this part of America, and traveled along the coast "with

---

[30]Oviedo, *Historia General,* II, 150. The late E. L. Stevenson had a photograph of a 250-page Chaves original manuscript which he found in the king's library at Madrid. I saw extracts from Chapter VIII of this manuscript and noticed little difference between the text and that published in the *Historia*. Stevenson thought it probable that Oviedo had at least some parts of the Chaves manuscript before him when he wrote. Oviedo himself asserts that he used an autograph Chaves map, II, 116, for his description of the American coast line. Cf. I. N. P. Stokes, *The Iconography of Manhattan Island, 1498-1909,* 6 volumes, New York, 1915-1928, II, 39-40.

a quadrant or an astrolabe in my hand . . . taking the latitude by the sun or by Polaris."[31]

The present writer reconstructed the Chaves map of the Gulf beginning with latitude 19°, according to the directions, distances, and latitudes given by Oviedo. The result is a sketch which closely resembles in outline that of Ribero.[32] With regard to the nomenclature from the Rio de las Palmas to the Rio de Nieves, we have a combination of the names which are found on the Weimar globe, and on the Ribero maps, plus four new names: "Rio Solo," "Rio de Pescadores," "Rio de las Montannas," and "Bahía del Espíritu Santo." According to Oviedo, Chaves inserted in the west coast of the Floridan peninsula a few new descriptive names and one geographical name: "B. de Miruelo." He had kept one legend, "B. de Juan Ponce," which is found on the four preceding maps; one, "R. de las Canoas," which is found on the Weimar globe and on the Ribero-Weimar of 1529, and one, "Rio de la Paz," which is found on the Weimar and on the Ribero-Borgia map. With regard to the position of the Rio del Espíritu Santo, we learn from Oviedo that on the Chaves map, the river emptied into a bay distant 160 leagues from Cabo Baxo, which is clearly the modern Cape San Blas in Florida. This means that the Rio del Espíritu Santo is in the vicinity of Galveston Bay, just as it is on the other maps.[33]

---

[31]*Historia General*, II, 136-137.

[32]The diagram from the Rio Pánuco to the Punta de Florida in Minnesota Historical *Collections*, VII, 1893, facing p. 28, has all the names along the north shore of the Gulf, only a few between the Rio Pánuco and the Rio del Espíritu Santo, and none on the west coast of the Floridan peninsula. The excerpt from Oviedo in *The Historical Magazine*, X, 1866, 371-374, is an adaptation rather than an English translation of the passage with quite a few errors of interpretation.

[33]Some years after the date of Chaves' map, Battista Agnese, a Genoese artist, began to produce many maps and atlases; cf. the extensive bibliography in H. R. Wagner, *The Manuscript Atlases of Battista Agnese*, a reprint for private circulation from *The Papers* of the Bibliographical Society of America, XXV, 1931. "The scientific merit of the Atlases of Agnese," says Harrisse, *Discovery of North America*, 628, "is very inferior to the artistic skill which they exhibit, . . . [they] possess no other merit than to be beautifully calligraphed or miniatured, not intended for use by mariners or scholars, and altogether objects of arts, destined to be offered as presents, and to adorn the libraries of the rich." Harrisse had said the

Despite the rhetoric about the Spaniards jealously guarding their maps and the oft-mentioned "obscurantism of the Church" which, we are told, was opposed to the diffusion of geographical knowledge, there were all over Europe, at this time, copies of maps based on the padrón which embodied the results of Spanish discoveries. This has been demonstrated by Harrisse and others. If additional proof were needed, it could be had from the fact that when in 1541 Gerard Mercator published the gores of his terrestrial globe at Louvain,[34] which was then in Spanish territory, he evidently had access to copies of maps similar to those made by the cosmographers of the Casa de Contratación.

If we examine the nomenclature of the Gulf Coast, we notice in the first place that limitations of space prevented the draughtsman or the engraver from inserting all the names which were on the map or maps at his disposal. Secondly, all place-names, with two exceptions, are to be found on maps previously analyzed; one of these new names is "R. de los angelos"; the other "Guadalquibil," is inserted between the Rio del Espíritu Santo and the Mar pequeña, and is also repeated on Mercator's map of 1569, but is not found on any previous or subsequent Spanish maps. Finally there are quite a few mistakes of transcription which show that the draughtsman or the engraver had an imperfect knowledge of Spanish.

The first and the third of the above points are so obvious as to need no elaboration; the second may be made clear from the following comparative list.

R. d palmas	Wolfenbüttel, Ribero - Weimar, Ribero - Borgia
Palmar	Chaves 1536
Cabo brauo	On all the maps
Costa tesa	Wolfenbüttel, Ribero-Weimar
Malabrigo	Wolfenbüttel, Ribero-Weimar
Magdalena	Weimar
[R.] de piscadores	Chaves 1536

---

same thing ten years earlier in *Jean et Sébastien Cabot. Leur origine et leurs voyages,* Paris, 1882, 189.

[34]Reproduced in facsimile after the original preserved in the Bibliothéque royale de Belgique, in *Sphère terrestre et Sphère céleste de Gerard Mercator de Ruppelmonde éditées à Louvain en 1541 et 1551,* Brussels, 1875.

R d gigâtes	Tierra de gigantes: Wolfenbüttel, two Ribero maps
C. d zierto	Chaves 1536
R del espiritu Santo	On all the maps
Guadalquibil	Cf. *supra*
Marpequena	On all the maps
Mata d Salmador	On all the maps
R. d Medam	Medanos: Weimar and the two Ribero maps
Costa d flores	R. de flores on all the maps
R de los angelos	Cf. *supra*
R de nieues	On all the maps
Nauidad	Wolfenbüttel
C. el Condido	*i.e.*, escondido: Wolfenbüttel
B. del guernador pâfilo	Cabot map
B. de vulguelo	*i.e.*, Miruelo: Chaves 1536

This shows that with regard to the nomenclature of the Gulf Coast, Mercator's globe closely follows maps based on the padrón. As for the position of the bay into which the Rio del Espíritu Santo empties, measurements show that on the earlier maps this bay is situated between 150 and 160 leagues from the west Florida coast. This distance can be computed by the simple process of applying the scale along the coast line. The Mercator map contains the longitudes as well as a scale. He says that each degree of longitude on the equator is equal to 18 Spanish leagues. According to this scale the bay is 144 leagues[35] from the west coast to Florida and its entrance is slightly below latitude 30°, which means that the bay is just east of present-day Galveston Bay, in approximately the same position which it occupies on the other maps.

In the year following the publication of Mercator's globe, Alonzo de Santa Cruz, one of the distinguished members of the geographical department of the Casa de Contratación, drew a map of the world.[36] The Gulf Coast as represented on this map

---

[35] The same distance is obtained if the longitudinal difference, between these two points, 9° 20′, is multiplied by the cosine of the latitude, 30°.

[36] Nova verior et integra totius orbis Descriptio nunc primum in lucem edita per Alfonsum de Santa Cruz Caesaris Caroli V. Archicosmographum A. D. M. D. XLII. Reproduction in phototypic facsimile . . . with explanations by E. W. Dahlgreen, under the title *Map of the World by*

is similar to that of the lost Chaves map of 1536, except that the exigencies of space forced the draughtsman to omit many place-names. In particular the position of the bay into which the Rio del Espíritu Santo is very nearly the same as on the Chaves map.[37] Sometime after 1560, Santa Cruz composed his *Islario General*.[38] The geographical names therein along the Gulf Coast are identical with those of the Chaves map of 1536, and the position of the bay is the same as on the other maps.

Sebastian Cabot's map of 1544 contains only slight variations from the Chaves map of 1536, with regard both to the nomenclature and to the position of the bay.[39] The Spanish names were copied by a man who was more familiar with Italian than with Castilian. The copyist's attention flagged at times; for instance, he has the Rio de Pescadores correctly placed, but where the Chaves map has Rio Escondido, we find "rio de pescado" instead. It must be observed that the course of the Rio del Espíritu Santo is indicated on this Cabot map as flowing directly south from latitude 37°, *i.e.,* the latitude of the mouth of the Ohio. It is obvious that if this river were the Mississippi, and if it had been explored as far as latitude 37°, the discoverers or explorers in their report to the geographers of the Casa would certainly not

---

Alonzo de Santa Cruz, 1542, Stockholm, 1892. In spite of what is said in the Latin title, the map was never published. Dahlgreen calls attention to the fact that on the Ribero map of 1529, an inscription at the mouth of the Congo River speaks of many converts to Christianity twenty years earlier. This inscription is repeated by Santa Cruz, but these conversions are said to have occurred thirty years ago; "thus it seems probable that, on his map of 1542 Santa Cruz copied without any change an inscription which he wrote down as early as 1539."

[37]At the latitude of the bay, one degree of longitude measures 15,155 Spanish sea leagues; and there are eleven degrees between the entrance of the bay and the west coast of the Floridan peninsula.

[38]*Islario General de Todas las Islas del Mundo,* A. Blázquez, ed., 2 volumes, Madrid, 1918.

[39]A photostat of the whole map was issued by the Massachusetts Historical Society in 1929. The map had been reproduced in facsimile by E. F. Jomard, *Les monuments de la géographie,* Paris, 1842-1862, nos. 64-71. Kretschmer, Taf. XVI, says that his reproduction is "nach Jomard," but it lacks the meridians and the parallels; a few errors of transcription were noted. Kretschmer's drawing is reproduced in Winship, *The Coronado Expedition,* between pp. 352-353.

have it emptying into such a large bay situated so far to the west. Hence we may conclude that the course of this river existed only in the mind of the mapmaker; and as we shall see, the same conclusion applies to many maps of the second group.

Before proceeding further it may be well to sum up the data obtained from the analysis of the maps already considered. Ribero, Santa Cruz, Chaves and Cabot were prominent members of the geographical department of the Casa; they all had access to the *Padrón General* and all drew their maps according to it, as they were bound to do under penalty of fine and even of loss of office. A comparative study of the section of the Gulf Coast under consideration shows that after 1529 very little change is found in the nomenclature from Apalachee Bay to Tampico, and no change at all in the position of the bay into which the Rio del Espíritu Santo empties; and the mouth of the river is always shown at the western end of the bay. These features, it must also be noted, appear on all maps after 1529, that is, after the ordinance of Charles V which ordered the revision of the padrón. Hence we conclude that from this time until 1544 there were no explorations of the Gulf Coast from Tampico to Apalachee Bay; or if there were, the reports of the pilots were not handed to the Casa. No matter which of these alternatives is chosen, it makes no difference concerning the question whether the Rio del Espíritu Santo was the Mississippi.

There are of course differences of detail in the numerous maps of the Gulf which have come down to us, but all that can be concluded from such differences is that the mapmakers of the Casa allowed themselves some latitude in copying the padrón. While careful to include all that would be useful to pilots, they did not draw all the maps exactly alike. It is quite certain that some lost maps of the technicians of Seville based on the padrón, though in substantial agreement with those which have come down to us, contained place-names not found on their extant maps.

This deduction is legitimate in view of the following facts. In 1541, that is, five years after Chaves' map and three years before the date inscribed on the Cabot map, Nicolas Desliens made, at Dieppe, a map of the world.[40] On it all the geographical

---

[40]Facsimile reproduction of the Western Hemisphere in Hantzsch and

names from Pánuco to the Mar pequeña are found as they appear on the early extant maps of the cartographers of the Casa which were analyzed in the preceding pages. Desliens' limited knowledge of Spanish enabled him to translate some names into French; Rio de Pescadores, for instance, becomes "R des pescheurs"; and Rio del Espíritu Santo is put down as "R. du S. esprit." Other names are Frenchified rather than translated, such as "C. brave" for C. Bravo; and there are mistakes of transcription such as "Montas de Salvador" instead of "Matas de Salvador." Next to the "Mar pequeña" there is a legend "C. Romo" not found on the earlier Spanish maps which have come down to us; and farther to the east, next to the "Coste de fleurs," the Rio de Flores of the Spanish maps, is the inscription "R. des angelz." Now these legends, "C. Romo" and "R. de los Angeles" appear on later maps made by Santa Cruz and Gutiérrez, two of the cosmographers of the Casa de Contratación. Hence we conclude that there were maps, now lost, based on the padrón as it appeared after 1536, which contained these names and which somehow reached Desliens. It cannot be seriously maintained that the Spaniards copied these names from the Desliens maps, because from 1527 on, the cosmographers of the Casa were ordered to modify the padrón only after having studied the reports of the pilots. The proof that Desliens is the copyist in this case rests on the fact that "C. Romo" is a descriptive name which he was unable to translate, just as he was unable to translate other descriptive names of the model.

Another indication that there were other maps made by the geographers of the Casa which differed in a few details from those extant today, and which were copied by non-Spanish mapmakers is to be derived from a consideration of the Desceliers map of 1550.[41] The model which he copied differed little from

---

Schmidt, *Kartographische Denkmäler,* Taf. II; in *Atlas brésilien,* (see bibliography), no. 5. The editors of the latter noted: "Nicolas Desliens, 1543 ou 1544 (quoique datée de 1541)"; the author of the map wrote on it "Faicte a Dieppe par Nicolas Desliens 1541."

[41][World map], "Faite a Arques Par Pierre Desceliers PBre: Lan 1550." British Museum, Add. Mss. 24065, reproduced in C. H. Coote, *Autotype Facsimiles of Three Mappemondes, . . . C. The mappemonde of Desceliers of 1550,* privately printed, [Aberdeen], 1898. Cf. A. Anthiaume,

that used by Desliens. It further appears that both these models were compilations from other maps, in so far as they reproduce the geographical and descriptive names found on extant maps of the cosmographers of the Casa. In the Desceliers map, just before the "B. de mirguelo," *i.e.*, Miruelo—which appears for the first time on the Chaves map of 1536—is found the legend "p° chico"; now this legend is not found on any extant map made by the geographers of the Casa until the so-called De Soto map.[42]

There do not seem to be any extant Portuguese copies of Spanish maps showing the Gulf of Mexico before 1550, although some were certainly made in Portugal before this date. Thus it is believed that the so-called Henry II map[43] and the Harleian[44]

Cartes marines, Constructions navales, Voyages de Découverte chez les Normands, 1500-1650, 2 volumes, Paris, 1916, I, 88-93.

[42]Cf. *infra*. What is said about the Desliens' and the Desceliers' maps is also true of "The Sey of the Indis of Occident" [Gulf of Mexico] by John Rotz, in his *Boke of Idrography,* British Museum, Old Royal Library, Ms. 20, E. IX, actual size photostat in the Edward E. Ayer Collection of the Newberry Library, Chicago. John Rotz, Jean Roze, a fellow townsman of Desliens and Desceliers, was the son of a Scot émigré, Ross. Still another Dieppois, Nicolas Vallard, composed an atlas in 1547; the original is in the Huntington Library, San Marino, California, photostat in the Ayer Collection. The map of "la neuue espaigne" is a compilation as are those of his Dieppe contemporaries. On Rotz and Vallard, cf. Anthiaume, *Cartes marines,* I, 61-75, 93-96.

[43][Mappemonde peinte sur parchemin par ordre de Henri II roi de France], facsimile in colors, in Jomard, *Les monuments de la géographie,* nos. 23-24; a part of it is reproduced in Kretschmer, Taf. XVII. The original was acquired by the Earl of Crawford, and was reproduced by Coote in his *Autotype Facsimiles, B.* The map is dated and signed: "Faictes à Arques par Pierre Desceliers, presb^re 1546." Anthiaume, *Cartes marines,* I, 86, writes as follows: "Pour expliquer la source des connaissances portuguaises et espagnoles utilisées dans la construction de leurs planisphères par Desceliers et d'autres cartographes dieppois, on a prétendu qu'en 1542 dom Miguel de Sylva, évêque de Viseu [cf. Winsor, *Narrative and Critical History of America,* II, 227], ayant été banni du Portugal, sa patrie, et qu'en venant en France, il emportait avec lui des cartes et des mappemondes qu'il livra aux Dieppois. Mais cette hypothèse est invraisemblable, ou tout au moins elle est insuffisante pour expliquer la science des Dieppois." The spelling of place-names which clearly indicates that the models used by the Dieppe cartographers were Portuguese versions of Spanish prototypes seems to have escaped Anthiaume. How valuable manuscripts—and undoubtedly maps also—passed from one country to another is illustrated in

were made after a Portuguese version of a Spanish map. The place-names from the Rio Pánuco to the Florida Keys are all found on Spanish maps, except three names on the west coast of Florida, just north of the Bahía de Juan Ponce: "C. escondido," "Rio Verde," and "Ascension." What led the Portuguese cartographers to insert these names, or what sources they had, is not ascertainable. It may be that outside of Spain mapmakers were freer, for in other countries there was no government control similar to that exercised by the officials of the hydrographic service of the Casa de Contratación.

Since all the maps of the first group were ultimately copies of Spanish prototypes which were based on the *Padrón General* in Seville, the nomenclature along the Gulf Coast is substantially the same on all of them, and on all these maps, the Rio del Espíritu Santo is shown emptying into a large bay located at the western end of the Gulf. Hence there is no necessity of recording here the results of our study of the numerous maps belonging to this group which were drawn or printed during the century and a half from 1520 to 1670.

Codex Mendoza, J. C. Clark, ed., 3 volumes, London, 1938, I, Introduction, 1-12. The ship carrying this manuscript to Spain was captured by the French, the *Codex* was acquired by André Thevet, from whom Hakluyt bought it.

[44][The Dieppe or Harleian World Map.] The original in the British Museum, Add. Mss. 5413, is reproduced in Coote's *Autotype Facsimiles, A;* the photograph in the Bibliothèque Nationale, Paris, Ge. DD 738 is taken from it. The map is dated "*circa* 1536" by Coote, it should be "after 1542," cf. Anthiaume, *Cartes marines,* I, 75-78.

III

THE SECOND GROUP OF MAPS

To the second group of maps belong those which show the course of the Río del Espíritu Santo long before there was any knowledge of it, if this river is the Mississippi. Thus on the already mentioned Cortés map, one of two large rivers flowing due south is called Río del Espíritu Santo, and three small rivers form the headwaters of the longer one, a hundred leagues from a twenty-league bay into which both large streams empty. It is manifest that this map, merely because it shows the course of these rivers, cannot be adduced as evidence that the Río del Espíritu Santo is the Mississippi. Obviously, the fact that the course of a river appears on a map does not prove that the river represented was actually explored. In the first place, we know that early mapmakers, except for such official cartographers as those of the Casa de Contratación, had many reasons for drawing the full course of rivers, though the models which they were copying represented only the mouth. A drawing of the entire course, however fanciful, would call attention to the hydrography of a given region much more emphatically than would a mere legend inserted near an indentation in the shore line. It would make the map "look better," and would even convey the suggestion that the map embodied the results of the latest discoveries. In fact, the titles of early maps sometimes make this suggestion explicitly to interest prospective buyers. Thus one of Allard's maps issued at the end of the seventeenth century is entitled: *Recentissima Novi Orbis sive Americae . . . Tabula,* although this "most recent" map is simply a reprint of another fifty years old. Moreover, if the appearance of a river's course on a map is taken as evidence that the river has been explored, we shall have to say that both the 500-mile rivers drawn on the Cortés map had been actually traversed by explorers from mouth to source by 1523. Finally, those who maintain that the shorter of these rivers, legended Río del Espíritu Santo, is intended to represent the Mississippi have to face the problem of identifying the second and longer river.

The portion of the Gulf shown on the Turin map[1] is a variant of that same portion on the Cortés map. The course of the Rio del Espíritu Santo as well as that of the longer river west of it is exactly the same on both. The place-names and their position on these two early maps of the Gulf differ as follows:

Cortés	Turin
Provincia amichel	[princia amichel]
[Tamacho pruvia]	tamacho princia
Rio panuco	Rio panuco
laoton	laotom
Tamacho pruvia	[tamacho princia]
[Provincia amichel]	Princia amichel
	rio de môtañas altas
R. la palma	rio de la palma
R. de Arboledas	
p° de arrecifos	p° de arrecife
Rio del Spiritu Sancto	rio del espiritu santo

Whereas on the Cortés map the coast line runs in an uninterrupted arc from the Rio del Espíritu Santo to the southernmost eastern tip of the Florida peninsula, on the Turin map the coast line stops east of the river, and after a gap of 175 leagues comes the "Isla Florida." Only three sides of this "island" are shown, and the position of its south shore is six degrees too far north.

The nomenclature of the Gulf Coast on the Paris Gilt Globe[2] is partly similar to that on the Cortés map, with the addition of latitudes and longitudes. The northern coast line is above the fortieth parallel, and the mouth of the "R. de S. Spiritu" is shown emptying into a bay at latitude 42°, the latitude of Chicago. Instead of being at the western end of the Gulf, this bay is located next to the Florida peninsula, where present-day Apalachee Bay is situated. The Rio Pánuco is represented as a huge stream flowing eastward. The Tamacho Provincia is on the northern bank of this river, as it is on the Cortés map; on the south bank, ten degrees from the Gulf Coast, we find Cathay (China). This globe expresses the conception of its maker rather than the actual

---

[1] Stevenson, *Maps Illustrating*, no. 6; facsimile in color in *Atlas brésilien*, no. 2.

[2] Nova et Integra Universi Orb[i]s Descriptio, in G. Marcel, *Reproductions de cartes et de globes relatifs à la découverte de l'Amérique du XVIe au XVIIe siècle*, avec le texte explicatif, Paris, 1892, *Atlas*, pl. 21. The changes on the "facsimile" in the *Discovery of North America*, 562, are inexcusable.

geography of the Gulf, and is quite useless as far as the identification of the Rio del Espíritu Santo is concerned.

The double-cordiform globe of Finaeus belongs to the same category.[3] As on the Paris Gilt Globe, America is joined to Asia forming a single continent. The nomenclature is that of the Cortés map. The bay into which the "R. de S. Spu" empties has its entrance above latitude 40°, and is situated as on the Paris Gilt Globe. The river is drawn north-south for ten degrees, beginning from James Bay; there the river forks, and the headwaters of the two branches, about five degrees north of the latitude of present-day James Bay, are four hundred miles apart.

The geography of the Gulf on this map is another example of the manner in which even professional geographers, like Finaeus,[4] relied on their own idea rather than on data supplied by accounts of voyages. Even if the latitudes were all lowered by ten degrees, and if the coast line were thus brought near its true position, the portion of the river represented would extend as far north as Chicago, and it is certain that no exploration of any river emptying into the Gulf of Mexico had been extended so far at this time.

The dozen place-names along the Gulf Coast on the Ulpius Globe[5] are taken from maps based on those of the cosmographers of the Casa de Contratación. The course of the "R. de S. Spu" extends over about five degrees, with its mouth at the northwest end of a bay situated in the northwest corner of the Gulf. A second stream of equal length empties into the same bay at its northeast end, and is called "R. Gvadalqvibir." As we have seen the earliest dated map on which this latter name appears is that of Mercator of 1541. The course of a great river which empties into the same bay at the same point is shown on the Vopel map in Girava,[6] but no name is given to this river.

---

[3]Nova, et Integra Universi Orbis Descriptio, in N. A. E. Nordenskiöld, *Facsimile-Atlas to the early History of Cartography,* Stockholm, 1889, pl. XLI; a part of this globe reduced to Mercator's projection is in Winsor, *Narrative and Critical History of America,* III, 11. In 1538, Mercator copied Finaeus' globe, cf. Nordenskiöld, *op. cit.,* 90, and pl. XLIII, photographic reproduction in *Revista de Indias,* Año II, 1941, no. 4, between pp. 100-101.
[4]Cf. Gallois, *De Orontio Finaeo Gallico Geographo,* Paris, 1890.
[5]Reproduced in Winship, *The Coronado Expedition,* facing p. 349.
[6]Typo de la Carta Cosmographica de Gaspar Vopellio Medeburgense,

An undated Agnese map also belongs to this group.[7] The early Spanish maps which he had previously copied contain only the names and mouths of various rivers. On this one, besides using the nomenclature of these maps, Agnese represented some of these rivers as arising in the foothills of a huge mountain range which runs in an immense arc from a point in the east between latitudes 35° and 40° toward the southwest and then south past Mexico City, from which point it runs in a southeasterly direction as far as the Bay of Campeche. One of these rivers is the Rio del Espíritu Santo, which is represented as flowing due south for 400 Italian miles, almost six degrees of latitude, until it empties into a bay called "B. del Spirito S^{to} della Colata." The mistaken name given to this bay has already been discussed.

This Agnese map, by the way, serves to illustrate the point which we have already noted; namely, that the appearance of the course of a large river marked Rio del Espíritu Santo on a map is no proof that this river is identical with the Mississippi. For although the Mississippi is admittedly the largest river that empties into the Gulf, the Rio del Espíritu Santo as drawn by Agnese is no larger than the four others which he represents as emptying into the Gulf along its northern coast line.

Notwithstanding its artistic merits, the map attributed to Jacopo Gastaldi and first published in Venice in 1554,[8] cannot be taken as a reliable representation of the nomenclature of the Gulf Coast nor of the hydrography of the interior. The river which occupies the position of the Rio del Espíritu Santo is left nameless and the bay into which it empties is called "Baia de todos Sanctos." The "R. de Canoas" which is located near the western end of the Florida peninsula on the early Spanish maps is here shown

---

in Hieronymo Girava, *Dos Libros de Cosmographia,* Milan, 1556; reproduced in Nordenskiöld, *Facsimile-Atlas,* pl. XLV.

[7]*Supra,* p. 2, note 3.

[8]Reproduced by F. Muller and Co., eds., in *Remarkable Maps of the XVth, XVIth & XVIIth Centuries reproduced in their original size,* 6 parts in 4 volumes, Amsterdam, 1894-1897, I, pl. 1-4; also in *Atlas brésilien,* no. 10a. The question of authorship is discussed by Nordenskiöld, *Periplus,* 146-147. H. R. Wagner, *Cartography of the Northwest Coast of America to the Year 1800,* 2 volumes, paged continuously, Berkeley, California, 1937, 279, lists it under Tramezini, the name of the printer.

as a large river, much longer than the Rio del Espíritu Santo, ending in Apalachee Bay.

The source of the nomenclature of the interior on the map engraved at Amsterdam in 1562 by Jerome Cock is discussed below, as are also the distortions which place-names underwent.[9] The hydrography is open to the same criticisms as is that of the maps previously described. The course of the Rio del Espíritu Santo is drawn, but other rivers east of it are also drawn, the longest of which empties west of Apalachee Bay, while the second longest has its mouth east of this bay and is called "R. de Juan Ponce." This latter name is evidently borrowed from a bay near the lower west end of the Florida peninsula, and no river of the same name appears on any map prior to this date. The arbitrariness of the engraver is all the more noticeable because he keeps the "Bª de Juâ ponce" in its traditional place between the 26th and the 27th parallel.

The Italian maps published between 1560 and 1570, as for instance those engraved by Berteli[10] and by Zaltieri,[11] are simply artistic fantasies. For all that pertains to the nomenclature of the Gulf Coast and the hydrography of the region north of it Zaltieri's map is a cartographical freak. None of these maps contains any reliable information whatever, and one cannot help wondering what model the draughtsman had before him when he was delineating the Gulf, and what was the purpose of issuing such bizarre productions.

The maps of the Portuguese cartographer Fernâo Vaz Dourado,[12] insofar as the nomenclature of the Gulf of Mexico is concerned, are all based on variants of the Spanish padrón. On

---

[9]Americae sive quartae orbis partis nova et exactissima descriptio; *Atlas brésilien,* no. 8. The map is dedicated to Margaret of Austria, Duchess of Parma, the regent of The Netherlands from 1559 to 1567.

[10]One dated *circa* 1560, has no title and is reproduced in *Remarkable Maps,* I, pl. 10; the other is entitled: Universale Descrittione di tutta la Terra Conosciuta fin qui . . . Paulo Veronese fecit. Ferando berteli Exc 1565, *ibid.,* IV, pl. 3.

[11]Original in the E. Ayer Collection, reproduced in Nordenskiöld, *Facsimile-Atlas,* 129, and in Winsor, *Narrative and Critical History of America,* II, 451.

[12]On Vaz Dourado, cf. Cortesão, *Cartografia e cartógrafos portugueses,* II, 7-28.

the earliest of these maps,[13] the "Rio do espirito Samto" disembogues in a bay at latitude 30°; there is a fork in the river three and a half degrees farther north, the two branches of which extend to latitude 37°, the top of the map. On another of his maps,[14] the same feature reappears: one branch comes directly from the north, the other from the west-northwest. The latter branch flows out of a large lake into which empty seven rivers. At about latitude 37°, the north branch receives a tributary coming from the east. Another large but unnamed river coming from the northeast also ends in this bay. This map also shows the course of other rivers which empty into the Gulf, though only their mouths are indicated on Vaz Dourado's previous maps. His map of 1571,[15] with its network of rivers, tributaries, and subtributaries, looks more like an arabesque than like an actual map. The fact that on two subsequent maps[16] Vaz Dourado did not

---

[13]Univercalis et Imtegra Totius Orbis. Hidrographia Adversisimam [sic] Lvzitanorum. Traditionem. Descriptio. Ferdinand Waz Dovrado. In Cevitate Goa. 1568. This atlas, in the Library of the Duke of Alva, Palacio de Liria, Madrid, is described by Cortesão, *op. cit.*, II, 28-41; the map of the Gulf of Mexico is the ninth, and is reproduced in color in *Atlas brésilien*, second memoir, no. 3. A photograph of the original is in the Karpinski Collection.

[14]The atlas, in which the map of the Gulf of Mexico is the second, is described by Cortesão, *op. cit.*, II, 68-77, and is in the Biblioteca Nacional, Lisbon, reproduced in color in *Atlas brésilien*, no. 18a; photograph of the original in the Karpinski Collection.

[15]On the restored title of this atlas, cf. Cortesão, *op. cit.*, II, 41-54. The map of the Gulf is reproduced in color in *Atlas brésilien*, no. 22a, a black and white photograph of it in Cortesão, pl. XLII; partial reproduction of the original in the Karpinski Collection.

[16]Universalis et Integra Totius Orbis Hidrografia. Ad Verissimam. Lvzitanorvm. Tradicionem Descripcio. Ferdinãdo Vaz. Este livro. Fes. Fernão Vãz Dovrado, British Museum, Add. Mss. 31317, actual size photostat in the Ayer Collection, described in Cortesão, *op. cit.*, II, 64-68. The Gulf of Mexico is the third map of the atlas; the title is written in a band around the map: "Nesta lamina esta lamcado todas as Amtilhas de Castela e nova Espanha ate a Florida; reproduced in color in *Atlas brésilien*, second memoir, no. 4.—The other map of the Gulf by Vaz Dourado is in an atlas of 1580 preserved in Munich, reproduced in F. Kunstmann, K. von Spruner, and G. M. Thomas, eds., *Atlas zur Entdeckungsgeschichte Amerikas,* Munich, 1859, Bl. X, and in *Atlas brésilien*, no. 26b. In his description of this atlas, Cortesão, *op. cit.*, II, 58, pertinently remarks: "Embora todos os seus

draw the course of the rivers at all may be taken as an indication of the importance which he himself attached to the hydrography of the earlier ones.

Mercator's map of 1569 also belongs to this second group.[17] Because this planisphere is the first example of his famous projection, it is a landmark in the history of cartography; but, as we shall see, the contribution which it made to a knowledge of the geography of the Southern United States is negligible.

An opposite view is expressed by Winsor, according to whom Mercator "was the first to map out a great interior valley to the continent, separated from the Atlantic slope by a mountainous range that could well stand for the Alleghanies. Dr. Kohl suggests that Mercator might have surmised this eastern watershed of the great continent by studying the report of De Soto in his passage to the Mississippi. . . ."[18] As a matter of fact the geographical feature mentioned by Winsor did not originate with Mercator. It is found on the maps of Cock and Zaltieri already mentioned; all that Mercator did was to delineate the mountain range more sharply than his predecessors. This mountain range may be intended to represent the Alleghanies, but it should be noted that the geographer shows it coming from the north and turning first west at the 35th parallel of latitude, along which it runs for twenty-five degrees, then due south to latitude 28°; where it turns at a right angle to the east, sloping toward the Tropic of Cancer and ending at longitude 269°. The Southern United States is therefore represented as an immense amphi-

---

trabalhos obedeçam a un tipo inconfudível, parece que Vaz Dourado tinha un Atlas protótipo, cuja parte meremente cartográfica foi aproveitando pera todos os outros, mas acrescentando-os e modificando-os conforme os novos conhecimentos geográficos adquiridos ou o fim a que os destinava."

[17]Nova et aucta orbis terrae descriptio ad usum navigantium emendaté accomodata, in *Drei Karten von Gerhard Mercator. Europa—Britische Inseln—Weltgarte.* Facsimile-Lichtdruck nach den originalen der Stadbibliothek zu Breslau. Hergestellt von der Reichsdruckerei. Herausgegeben von der Gesellschaft für Erdkunde zu Berlin, London, Berlin, Paris, 1891, 3 parts in 1 volume; Bl. 8 of the Weltkarte, third part, has the Gulf of Mexico; partial reproduction of this sheet in Winship, *The Coronado Expedition,* between pp. 376-377.—*Infra,* pl. 4. (II)

[18]J. Winsor, *Cartier to Frontenac. Geographical Discovery in the Interior of North America in its Historical Relations 1534-1700,* Boston and New York, 1894, 66.

theater completely shut off from the St. Lawrence basin by this range of mountains, so that the only approach of it by land is from Mexico. This conception was taken over by subsequent mapmakers, and its influence is manifest among theoretical geographers and even among actual explorers of this region for more than a century after the publication of Mercator's map.

In view of Mercator's dependence on Cock, which is especially noticeable in his use of identical nomenclature, the position of Winsor and Kohl is untenable. The only report of the De Soto expedition which Mercator could have known is the book of the Gentleman of Elvas, which had been in print twelve years at the time when the 1569 map was published. The source utilized by Mercator, however, was not this Portuguese narrative, but the map of Cock, as can be seen from the following table. The words italicized are clearly indicated on both maps as being names of towns,[19] the names in roman are those of provinces or regions.

Cock 1562	Mercator 1569
Mocosa	Mocosa
Avacal	Auacal
Canagali	*Canagali*
Coruco	*Coruco*
Apalchen	Apalchen
Otagil	Tagil
Tierra Florida	La Florida
Iguas	*Iguas*
Calicuaz	Calicuaz
Comos	*Comos*
Cossa	Cossa
Capaschi	Capaschi
Monte Suala	

Such a perfect parallelism can hardly be a coincidence. The sources from which Cock derived this nomenclature will be ascertained below. We may here observe in passing that some of the names are recognizably identical with those which were first mentioned by the chroniclers of the De Soto expedition. Thus Mocosa is the Mocoço of Ranjel and Elvas; Apalchen is Apalache; Cossa is spelled Coça in Ranjel and Elvas, Coza in Biedma's chronicle, and Cossa on the so-called De Soto map; Monte Suala

---

[19] The names of the four towns are also on Zaltieri's map previously mentioned.

on the Cock map is the Xuala or Xualla of the chroniclers. In the case of other names, the distortion is too great to permit of more than probable identification. Thus Otagil, Tagil may be standing for Agile or Axile; Calicuaz for Coligoa or Coligua, Capaschi for Capachequi.

Another indication that Mercator's planisphere was compiled from maps then available is the following. Six degrees west of the Texas coast, on the northern bank of the Rio Pánuco, there is a town named Colaoton. In a previous section we called attention to the fact that the German engraver of the Cortés map inscribed "Rio panu-colaoton" on his map, and that the Italian engraver of the same map mistakenly read "colaoton" for "laoton" as the name of the region on the banks of the Rio Pánuco. On Mercator's planisphere the same mistake is repeated, and besides, the region is changed into a town 350 miles inland. It cannot even be said that this last mistake was original on Mercator's part, for he merely copied this legend from Zaltieri's map.

The similarities pointed out in the preceding paragraphs leave little doubt as to the sources utilized by Mercator. We must now attempt to ascertain from what source Cock himself obtained the inland nomenclature north of the Gulf of Mexico. He could, of course, have made direct use of the narrative of the Gentleman of Elvas, the only chronicler of De Soto's expedition whose work was then in print, but considering his distortions of the names mentioned in Elvas, some of which we have noted in connection with the list given above, it is more likely that his actual source was a model map on which these names were misspelled or else so badly written that he had to guess at their spelling.

This presumption is seemingly borne out by the title of Cock's engraving as catalogued by the British Museum, where the engraving itself is to be found: "Americae, sive quartae orbis partis nova et exactissima descriptio. Auctore D. Gutiero. 1562."[20] This author of the original was Diego Gutiérrez, a member of the geographic department of the Casa de Contratación, about whom little is known.[21] Few of his maps have come down to us

---

[20]*Catalogue of the Printed Maps, Plans, and Charts in the British Museum,* London, 1885, col. 1704, 69.810 (18).

[21]There were two Diego Gutiérrez, father and son. The author of this map is more probably the son. Cf. Fernández de Navarrete, *Biblioteca*

in their original form. One, signed and dated 1550, does not have any inland nomenclature at all.[22] The original of this map, which is in the Archives du Service Hydrographique, Paris, may have been made on an earlier model, otherwise we cannot readily understand how a map dated 1550 should lack nomenclature that was certainly known at this date by the cosmographers of the Casa. Now the coastal nomenclature which is found on Gutiérrez' map of 1550 is so difficult to decipher that many place-names can be identified only by their more legible appearance on the other early Spanish maps which were ultimately based on the same matrix as that of Gutiérrez, namely, the padrón in Seville. The Gutiérrez map from which Cock made his 1562 engraving may well have contained the inland nomenclature which does not appear on the 1550 map preserved in Paris; but these place-names would very likely have been so difficult to decipher that the distorted form given by Cock to the De Soto nomenclature is readily understandable. It is manifestly impossible to reconstruct the original map from Cock's engraving of it, for we have already seen an instance of his arbitrary way of altering the map which he engraved, and it is to be feared that he took similar liberties with the nomenclature of the interior as well as with the position of place-names on the map of Gutiérrez.

Thus the source of the orography and nomenclature of the Southern United States on the Mercator planisphere of 1569 is traced back, through Cock's map, to a map made by a member of the Casa de Contratación at some unknown date after 1550. With regard to the nomenclature along the coast, Mercator used his own map of 1541 as a basis, adding some place-names which are found on the earlier maps discussed in the previous section. That the hydrography of the same region was not derived from Cock or the reports of the De Soto expedition is clear from the following consideration.

Between the Rio Pánuco and the Rio de Perla, which is half-

Marítima Española, 2 volumes, Madrid, 1851, I, 342-343; Marcel, *Reproduction de cartes et de globes*, text, 108-109; Harrisse, *Jean et Sébastien Cabot*, 231-236.

[22]Diego gutierrès cosmographo de Su mag.ᵈ Me fizo enseuilla Año de 1550; ASH, 116-1, in Marcel, *Reproductions de cartes et de globes*, pl. 31-34. The reduction in the Karpinski Collection is too great to be of use.

way down the west coast of Florida, Mercator's map represents the course of ten rivers, six of which empty into the Gulf on the north. Of these latter, the Rio de Pescadores is the longest; the Rio del Espíritu Santo, the second longest, empties not directly into the Gulf itself, but into a "Baia de Culata," which occupies the position of our modern Galveston Bay, twelve degrees west of the Gulf Coast of Florida. A comparison with Cock's engraving will show that these rivers as drawn by Mercator were not copied from Cock's, and an examination of the reports of the De Soto expedition proves that Mercator did not make use of these reports, because the rivers on his map are neither named nor described therein.

In particular, it is impossible to maintain that the Rio del Espíritu Santo on the map of 1569 is the Mississippi. For although the latter river had been explored by the De Soto expedition twenty-five years before the date of this map, yet the Rio del Espíritu Santo is not represented as the largest river emptying into the Gulf, but rather as smaller and shorter than three other rivers, namely the Rio de Pescadores, the "Palmar r." and the "r. Solo."

Mercator's map was widely imitated. Thus in 1570, a simplification of it appeared in Ortelius' atlas, but the exigencies of space prevented the draughtsman or the engraver from inscribing the names of the rivers along the coast. A map of the Western hemisphere[23] which appears in the same atlas includes a greater number of legends from the planisphere of 1569, but again the size of this map prevented the inclusion of all the geographical details found on the model. Certain changes of positions are purely arbitrary, and cannot be the result of a better understanding of accounts of travel, or of additional information about the country represented.

Notwithstanding Thevet's grandiloquent assertions to the contrary, the map of the Western hemisphere[24] in his *Cosmographie*

---

[23] Americae sive Novi Orbis, Nova Descriptio, in *Theatrum Orbis Terrarum*, Antwerp, 1570. In the list of maps at the end of this essay, those which are mere variations of the simplified Mercator planisphere are classified as (IIa).

[24] Le Novveav Monde Descovvert et Illustre de Nostre Temps, in *La Cosmographie Universelle D'André Thevet Cosmographe du Roy*, Paris, 1575.

Universelle is purely and simply a copy of Mercator's map. Some of the latter's Spanish place-names are translated into French, while others, as in the case of Desliens, are Frenchified rather than translated. Many of the legends in the text of Thevet are copied from Mercator's planisphere, but are not inscribed in his map. On the east bank of the "r. d S. Esprit," Thevet has two small tributaries which are also marked on the map of the Western hemisphere of Rumold Mercator,[25] and which have been identified by imaginative writers as the Ohio and the Illinois rivers! This map of 1587 became even better known than the 1569 planisphere because it was reproduced in the numerous editions of Mercator's atlas. The map in Jode's atlas, dated 1589,[26] as well as Michael Mercator's map of 1611,[27] and many others either in print or in manuscript are all ultimately based on the planisphere of 1569.

The manuscript maps of Martínez which are not clearly based on the Mercator model but yet show the course of some rivers emptying into the Gulf require special mention here.[28] They show

---

[25]*Orbis Terrae Compendiosa Descriptio Quam ex Magna Vniversali Gerardi Mercatoris . . . Rumoldus Mercator fieri curabat A° M.D. LXXXVII*, reproduced in Nordenskiöld, *Facsimile-Atlas,* pl. XLVII, in *Atlas brésilien,* no. 33, in Winship, *The Coronado Expedition,* between pp. 388-389, etc.

[26]Totius Orbis Cogniti Universalis Descriptio, in *Speculum Orbis Terrarum,* [Antwerp, 1593-1613].

[27]*America sive India Nova, ad Magnae Gerardi Mercatoris avi Vniversalis imitationem in compendium redacta. Per Michaelem Mercatorem Duysburgensem,* in the fourth edition of Mercator's atlas published at Amsterdam by Jodocus Hondius. The date, 1611, was changed into 1616.

[28]"The Gulf of Mexico and the Caribbean Sea, with the eastern Coast of America from 46° north to the line," title from the *Catalogue of the Manuscript Maps, Charts, and Plans, . . . in the British Museum,* 3 volumes, 1844, I, 29. This map is no. 14 of an atlas of eighteen manuscript maps, Harleian Ms. 3450, it is reproduced in the Karpinski Collection, actual size photostat in the Ayer Collection. Cf. the maps of a section of North America and of the Gulf of Mexico in a Martínez atlas of 1587, Madrid, Bib. Nac., CC 35 ER 6. nos. 14 and 16.—There is in the Biblioteca Nacional of Madrid a 1587 atlas by Chretien Sgrooten. The nomenclature is that of the Mercator planisphere, the rivers are all giant streams; the largest of all empties in the northwest corner of the Gulf and is called Rio de Piscadores. Its headwaters are at latitude 42°, in the foothills of a mountain range which separates the Floridan basin from the headwaters of subtributaries of the St. Lawrence.

the influence of Mercator insofar as the nomenclature of the provinces and towns of the country north of the Gulf is concerned, but the place-names along the coast line which are derived from early Spanish maps are more numerous than on the maps listed in the preceding paragraphs. As for the course of the rivers represented by Martínez, it is simply an exercise of imagination.

In the Jode atlas referred to above, there is another map of the northern hemisphere which differs from the planisphere in several particulars.[29] The most notable difference is that just west of the Rio del Espíritu Santo, there is a much longer and wider river called R. del Oro. One degree from its mouth this river divides itself into two branches with their headwaters 500 miles to the north in the foothills of the mountain range which encircles Florida. Its appearance is not unlike that of the Rio del Espíritu Santo on the Cortés map.

The above list of maps belonging to the second group is far from exhaustive, but nothing would be gained by lengthening it. From the examples given it should be clear that the appearance of the course of a river on a map does not mean that the mapmaker entered it after studying accounts of travelers, nor does it mean that the river has been explored. All that can be deduced from an examination of the maps of the second group is that on many of them a river named Rio del Espíritu Santo empties into a large bay in the northwestern corner of the Gulf. We have seen that these maps are based on a few models, mainly the Cortés map of 1524 and the Mercator planisphere of 1569, and an analysis of these model maps makes it quite clear that the course of the rivers shown on them has little more foundation than the fancy of the cartographer. Consequently, the maps of this second group do not prove the identity of the Rio del Espíritu Santo with the Mississippi any more than do the maps of the first group. First, there is the bay into which the river empties. Geologists assure us that the Mississippi in historic times did not empty into a bay, but directly into the Gulf. Secondly, the position of this bay is shown on all these maps much farther west than the actual position of the mouth of the Mississippi. These arguments

---

[29]Hemispheriũ ab Aequinoctiali Linea, ad cirulũ Poli Arctici; also reproduced in *Atlas brésilien,* no. 35.

have already been presented in our discussion of the first group of maps, but they apply here as well, because the coast line, as shown in the second group is based on the same model as the first group, namely, the padrón in Seville. Finally there is admittedly no river emptying into the Gulf which is comparable in size with the Mississippi, and yet on many of these maps other rivers are represented as longer and wider than the Mississippi actually is.

IV

THE NARRATIVE OF ALVAR NÚÑEZ CABEZA DE VACA

Before studying the third group of maps we shall briefly examine Alvar Núñez Cabeza de Vaca's account of his journey along the Gulf Coast from St. Mark's Bay to Galveston. This narrative is worthy of detailed study because it is the earliest account of an exploratory voyage along this section of the coast, and remained the only account for more than a century and a half.

In June 1527, Pamphilo de Narváez left Spain for America. This expedition is recorded on the Weimar map of that year as the following legend on the map indicates: "Tierra que aora va apoblar panfilo de Narbaes." The earliest extant Spanish map which indicates the place where Narváez landed is the Cabot map, whereon the following note is inscribed near the legend "baya de miruelo" [Apalachee Bay], "Aqui desambarco panf¹º de Narvaez." This map is dated 1544, seven years after Cabeza de Vaca had returned to Spain with the news of the fate of the expedition. It is impossible to say whether the geographers of the Casa received a written report similar to that prepared by Cabeza de Vaca, Dorantes and Castillo for the Audiencia at Santo Domingo, or whether they used Cabeza de Vaca's book which was published in 1542, two years before the date of the Cabot map. It may be that the Spanish cartographers did not take Cabeza's report very seriously, for as has been observed, "there are few Spanish narratives that are more unsatisfactory to deal with by reason of the lack of directions, distances and other details, than that of Cabeza de Vaca."[1] Although the author of this comment is referring primarily to the route followed across the continent from Texas to Sonora, the Cabeza narrative is just as unsatisfactory with regard to directions, distances, and other details along the Gulf Coast.

As for the relative merits of the letter to the Audiencia and

---

[1] F. W. Hodge and T. H. Lewis, eds., *Spanish Explorers of the Southern United States 1528-1543*, New York, 1907, 7. This volume comprises three narratives, that of Cabeza de Vaca, that of the Gentleman of Elvas, and Castañeda's narrative of the Coronado expedition. Hodge edited and annotated the first and the third, Lewis the second.

Cabeza's book of 1542, Bandelier says: "Oviedo, who gives the text in full of the letter handed to the Audiencia of Santo Domingo by Cabeza de Vaca and Castillo when they touched that port on their return to Spain in 1537, has used the 1542 print for comparison with that letter . . . [and] inclines in favor of the former."[2] He adds that the two accounts do not conflict "on important points" and that "on the whole the difference between the two documents is so slight that there has been no occasion to publish [in English] the Letter to the Audiencia also." In this letter, however, there are important details regarding the location of the Bahía del Espíritu Santo into which the river of the same name emptied, which are not found in the book and which bear on the question at issue in this essay.

Bandelier further remarks: "A very serious objection to the credibility of the three narratives,[3] arises from the fact that all are based upon recollections only, and not upon journals or field-notes of any kind. It was, of course, impossible for the outcasts, shifted and shifting from tribe to tribe, to keep any written record of their trip."[4]

Information of this sort would hardly be used by the official cosmographers for the padrón, as we can see from the following note which López de Gómara inserted at the end of his description of the American coast in his *Historia General de las Indias:* "The number of leagues and the latitudes set down by me are according to the maps of the royal cosmographers, who neither receive nor accept the report of any pilot except under oath and [when cor-

---

[2] F. Bandelier, translator, *The Journey of Alvar Núñez Cabeza de Vaca,* New York, 1905, introduction by A. Bandelier, xiv.

[3] Bandelier is referring to (1) the *Relacion que dio Aluar Nuñez Cabeça de Vaca* . . . , Zamora, 1542, reprinted at Valladolid in 1555, under the title *La relacion y comentarios del gouernador Aluar Nuñez Cabeça de Vaca* . . . ; the *Comentarios* added to this edition are by Pero Hernandez, and relate to Cabeza de Vaca's career in South America. I have been unable to consult the 1542 edition, but according to Hodge the differences between the two texts are slight. (2) The letter to the Audiencia of Santo Domingo, in Oviedo, III, 582 ff.; this is an earlier account than the preceding one. (3) The "Relación de Cabeza de Vaca, tesorero que fué en la conquista," in *Colección de documentos inéditos,* XIV, 269-279.

[4] Bandelier, *op. cit.,* introduction, xvi.

roborated by the testimony of] witnesses."⁵ From the historical legend inscribed on the Cabot map, we know that the journey of Cabeza de Vaca was known to the geographers of the Casa; they certainly had the book and they may also have had a copy of the collective letter to the Audiencia of Santo Domingo. Yet the only legend which distinctly refers to the Narváez expedition is this one on the Cabot map, which may never have been inserted on the official padrón at all. The attitude of the Casa toward all this information may be gathered from the following comments of Oviedo:

> But in some way, I consider more trustworthy and clearer the relation of the three [the letter to the Audiencia] than the other [scl. the book] which only one wrote and which he published; as I say, I am taking from the [collective] letter and from what Cabeza de Vaca himself [told me] what is added to this chapter. They express themselves clearly and say what needed to be said, in spite of the fact that, as we have seen, having undergone such hardships they did not make a formal report or kept a record of their route, or through what latitudes their aimless wanderings took them. . . . I am not surprised at this, since the pilot himself, Miruello by name, was unable to guide them to the port which the fleet set out to find, and he was unable to tell them where he landed them or where they were. . . .⁶

Moreover, the very position of the legend on the Cabot map⁷

---

⁵López de Gómara, *Primera y segunda parte de la historia general de las Indias,* Fol. ix verso.

⁶*Historia General y Natural de las Indias* III, 615.

⁷"The Bahia de la Cruz of Narvaez' landing made identical with Apalachee Bay by Cabot, is likely to have been by him correctly identified, as the point could be fixed by the pilots who returned with the ships to Cuba, and would naturally be recorded on the charts. . . . Narvaez and his party evidently thought that they were nearer Pánuco, and had no idea they were so near Havana. Had they been at Tampa Bay, or on a coast running north and south, they can scarcely be supposed to have been so egregiously mistaken," Shea, in Winsor, *Narrative and Critical History of America,* II, 288. With regard to the position, Cabeza de Vaca distinctly states that Miruelo and the other pilots had no idea as to where they were. López de Gómara gave the conditions which were to be fulfilled before new discoveries were inserted in the official map.

It is important to bear in mind that the maps which have come down to us are not pilot maps. For the most part they are artistic productions

which marks the supposed landing place of Narváez is so far from accurate that it must have been based on conjecture rather than on actual sworn data, and this is a further reason why the geographers would not insert it in the padrón. Most modern writers locate Narváez's landing place "at St. Clement's Point on the Peninsula west of Tampa Bay."[8] This identification is much too precise in view of the fragmentary and unsatisfactory evidence on which it is based, but we call attention to it because the same overprecision appears in modern accounts of Cabeza de Vaca's discovery of the Mississippi. It is readily admitted that coming from the east and keeping close to the coast, the makeshift boats must have passed the mouths of the Mississippi, but whether the usually quoted passage in Cabeza de Vaca's relation designates the Mississippi is far from certain.

Miruelo was boasting when he assured Narváez that he knew the northern coast of the Gulf and the way to the Rio de las Palmas. After landing, he did not know where he was, nor where was the port near the Rio de las Palmas whereto he had promised to steer the ships. Two weeks later, Narváez held a council which Cabeza de Vaca attended. The commander told them that he intended to march inland, while the ships would lounge the coast to the harbor which they believed was not far distant. This plan was opposed by Cabeza de Vaca, but Fray Juan Xuárez was all in favor of it. He advocated going in search of the harbor which the pilots said was not more than ten or fifteen leagues from where they were in the direction of Pánuco. By following the coast they could not fail to come upon this port, because, they believed, the sea entered the land a dozen leagues.[9] This port was the mouth of the Rio de las Palmas, whither Narváez first in-

---

that give a fairly accurate, but general idea of the geography of the known world. Between these maps and the sailing charts given to the pilots there was a difference similar to that which exists today between maps found in ordinary atlases and the maps issued by the United States Coast and Geodetic Survey or by the British Admiralty.

[8]W. Lowery, *The Spanish Settlements Within the Present Limits of the United States 1513-1561*, New York, 1901, 177, and Appendix J, 453-455, where the various opinions up to the time of publication are listed. Hodge, *Spanish Explorers of the Southern United States*, 19, note 1, seemingly accepts this identification.

[9]*Spanish Explorers of the Southern United States*, 22-23.

tended to go, and to which Miruelo had claimed to know the route. From what Xuárez is reported to have said, it is clear that the pilots were just as ignorant of the geography of the Gulf as Miruelo was, and it is little wonder that the geographers of the Casa would not enter in the padrón the "Bahía de la Cruz" where Narváez landed.

Returning from the Apalachee country, the expedition arrived at a bay called "Bahía de Caballos" by Cabeza de Vaca and which is identified as St. Mark's Bay.[10] They left this place in the makeshift boats, September 29, 1528, and followed the coast "in the direction of the Rio de las Palmas," that is, they followed the west shore of Apalachee Bay, which runs in a southwesterly direction to Cape San Blas. Here the shore line begins to run northwestward, but Cabeza de Vaca does not mention any change of direction during the thirty days they followed the coast to a place where they "found shelter with much calm."[11] This place has been identified as Pensacola Bay. After the rough weather had subsided they "again embarked and navigated three days,"[12] at the end of which they entered an estuary (estero). If the previous stopping place is Pensacola, the estuary should be Mobile Bay; commentators, however, omit to identify this place. After a brush with the Indians, the Spaniards put to sea followed by the natives hurling clubs and throwing stones at them "until mid-day."[13]

> We sailed that day[14] until the hour of vespers,[15] when my boat, which was leading, discovered a point made by the land, and against a cape opposite, could be seen a very great river.[16] I cast anchor near a little island forming the point, to await

---

[10]*Ibid.*, 37, note 2.
[11]*Ibid.*, 38, note 1.
[12]*Ibid.*, 39.
[13]*Ibid.*, 41. The Spanish text of the 1555 edition as well as the Bandelier translation of the 1542 edition of the *Relacion* were used to emend the translation published in *Spanish Explorers*.
[14]Hodge, *Spanish Explorers*, 41, note 1, observes that "according to the Letter [to the Audiencia] they travelled two days more before reaching this point of land."
[15]"Hasta hora de visperas," Bandelier rendered this "until nightfall"; Hodge in *Spanish Explorers*, "the middle of the afternoon." The time of vespers, according to the season, varied between 4:00 and 6:00 P. M.
[16]"Un río muy grande."

the arrival of the other boats. The Governor [Narváez] did not want to come there, but entered a bay near by in which were many islets. We came together there, and took fresh water from the sea, for the stream entered it in freshet.[17] To parch some of the maize we had brought with us, —we had eaten it raw for two days—we went on that island;[18] but finding no fire-wood, we agreed to go to the river beyond the point, one league off. By no effort could we get there, so strong was the current,[19] which drove us out, while we contended and strove to gain land. The north wind, which came from the shore, began to blow so strongly that it forced us to sea without our being able to overcome it. We sounded half a league out, and could not find bottom at thirty fathoms; but we were unable to satisfy ourselves whether it was the current that was the cause of our failure to find bottom.

Eight days after the adventure at the mouth of this river, Cabeza de Vaca and a few of his unfortunate companions were cast ashore on an island which is identified by some as Galveston Island, by others as Velasco Island,[20] just south of the former. The very great river with the strong current is "the Mississippi the waters of which were now seen by white men fourteen years before the 'discovery' of the stream by De Soto."[21] This identification is made by the same author who identified Pensacola Bay, which was reached after a navigation of thirty days from St. Marks' Bay, a distance of 200 miles along the coast. The unidentified estuary was reached in three days and the Mississippi in one more day, which means that it took them only four days to travel 250 miles, and only eight more days to cover the 350 miles between the mouth of the very great river with the strong current to Galveston.

Only when the chronology of Cabeza de Vaca's narrative between the intermediary points of the two termini, St. Marks' Bay and Galveston, is compared with the distances does the difficulty of identifying these intermediary points become manifest.

---

[17]"Porque el río entraua en la mar de auenida"; Bandelier's translation: "because the river emptied into it like a torrent."
[18]"Saltamos en aquella isla," namely, the island forming the point.
[19]"Era tanta la corriente."
[20]*Spanish Explorers*, 57, note 2.
[21]*Ibid.*, 41, note 2.

Even if we should take the time given by Oviedo—two days—between the estuary and the mouth of the very great river with the strong current, this would mean that they went from Pensacola Bay to the "Mississippi" in five or six days, or that the Spaniards in wretched boats, exhausted by fatigue and privations, managed to cover, in stormy weather, forty miles a day.

Neither the greatness of the river nor the strength of the current are of much use in identifying it, for Cabeza de Vaca uses the same words to describe rivers which are certainly not comparable with the Mississippi.

During the march in Florida, "we came upon a river which we crossed with very much trouble by swimming and on rafts. It took us a day to cross it because its current was very strong."[22] This river is "evidently the Withlacoochee which enters the Gulf at latitude 29°," according to Hodge.[23] "That night we reached a river which was very deep and very wide, and its current very swift."[24] This is identified as the Suwannee River.[25] Again a "very great river" which the Spaniards named Rio de la Magdalena[26] is said to be present-day St. Marks River.[27]

Cabeza de Vaca did not recognize the Mississippi—if that be the great river with the strong current—as the Rio del Espíritu Santo, although he knew that somewhere on the coast there was a river thus called which emptied into a bay of the same name, for both he and his companions looked for it much farther west than the mouth of the Mississippi. After meeting Dorantes, the survivors decided to remain on the island, but they "also agreed that four men of the most robust should go to Pánuco, which they believed to be near."[28] Dorantes, one of those who went in

---

[22]"Llegamos a vn rio que lo passamos con muy gran trabajo a nado y en balsas: detuuimonos vn dia en passarlo que traya *muy grã corriente.*" *La relacion y comentarios,* Fol. vii verso.

[23]*Spanish Explorers,* 25, note 3.

[24]"Aquella noche llegamos a vn río, el qual era *muy hõdo y muy ancho, y la corriente muy rezia,*" *La relacion,* Fol. viii verso.

[25]Hodge, *Spanish Explorers,* 27, note 1.

[26]"Ya en este camino la auiamos descubierto por vn *río muy grande* que en el hallamos, a quien auiamos puesto por nombre el río de la Magdalena." *La relacion,* Fol. ix verso.

[27]Hodge, *ibid.,* 33, note 1.

[28]*Spanish Explorers,* 49. Cf. the letter to the Audiencia, Oviedo, III, 591.

search of Pánuco, "followed the coast forty leagues farther," in a southwesterly direction and "passed three times by a bay (ancon), which he says he believes from its appearance to be the one which they call *del Espiritu Sancto*." After crossing another bay with the help of the Indians, "they marched two leagues to a great river which was beginning to swell owing to inundation and rains";[29] they made rafts and had much trouble in crossing it, because there were few swimmers among them. From this latter bay they marched three leagues to another very mighty and flooded river, which flowed "with such violence that fresh water went out very far into the sea."[30] Continuing their journey southward, the travelers crossed other "rios grandes," until they arrived at a bay which was "wide and about one league across; the side in the direction of Pánuco juts into the sea about one fourth of a league; it has huge hillocks of white sand, which must be seen far out in the sea, and for this reason they conjectured that it must be the *Rio del Espíritu Sancto*."[31]

From what has been said several conclusions appear legitimate. First, there is no reason for saying that "the river with the strong current which entered the sea in freshet" is the Mississippi either from its position on the Gulf coast, or from its description in the narratives. Secondly, there is no more reason for identifying the Mississippi as the great river with the strong current spoken of by Cabeza de Vaca than there is for identifying it with the very large and mighty river mentioned in the royal *cédula* of 1521. Finally, the narrative of Cabeza de Vaca certainly does not prove the identity of the Mississippi and the Rio del Espíritu Santo.

---

[29]"Hasta un rio grande que commençaba á crescer por avenidas é lluvias," Oviedo, III, 593.

[30]"Hasta otro rio que venia muy poderoso é avenido é con tanta furia que salia el agua dulçe muy grand rato en la mar," *ibid*.

[31]Oviedo, III, 593. Speaking of the terminus of this journey, Oviedo notes: "é que assi llegaron al ancon ques dicho que creian ques el del Espiritu Sancto," *ibid.*, 594. After leaving Malhado Island, Cabeza de Vaca crossed "el ancón y quatro ríos que ay por la costa, . . . y ansi, fuymos con algunos indios adelante, hasta que llegamos a vn ancon, que tiene vna legua de traués: y es por todas partes hõdo: y por lo que del paresció y vimos, es el que llamã *del spiritu sancto* . . . ," *La relación y comentarios*, Fol. xxii. See the variant in *Colleción de documentos inéditos*, XIV, 279.

V

THE NARRATIVES OF THE DE SOTO EXPEDITION—
THE THIRD GROUP OF MAPS

Cabeza de Vaca arrived in Spain while De Soto was preparing his expedition to Florida. The latter sought to enlist this only member of the Narváez expedition then in Europe, but Cabeza de Vaca, having plans of his own, declared that he could not give any information unless released from his oath of secrecy by Dorantes, who had remained in Mexico. To Charles V, however, he spoke in such a manner as to imply that Florida was the richest country in the world, and to some of his relatives he intimated that they would act wisely in accompanying De Soto.[1] What Cabeza de Vaca had seen of Florida hardly justified such a glowing account, and what he said must have been radically different from what we read in the relation he published five years later.

In this section we shall attempt to determine, by means of three of the four narratives of the De Soto expedition, those of Biedma, Ranjel and the Gentleman of Elvas,[2] and from an examination of two maps based on information contained therein, whether the Rio del Espíritu Santo is the Mississippi. The fourth narrative, that of Garcilaso de la Vega, will be studied in the next section because it was not published until sixty years after the return of the survivors, and because on this narrative is based a map representing a new concept of the geography and hydrography of what was then called Florida, *i.e.*, the Southern United States. We assume that the reader is familiar with the three accounts, with their provenance, their authenticity, and their evidential value.[3] We shall briefly recall what is known about their approximate date of composition.

---

[1] E. G. Bourne, ed., *Narratives of the Career of Hernando de Soto in the Conquest of Florida,* 2 volumes, New York, 1922³, Narrative by the Gentleman of Elvas, I, 5-9.

[2] For the sake of convenience we shall refer throughout this section to the translations in Bourne's *Narratives.* The bibliographical introduction is now superseded by that in the *Final Report* of the United States De Soto Expedition Commission, 76th Congress, 1st Session, House Document No. 71, Washington, D. C., 1939, 4-9.

[3] Cf. T. H. Lewis, "The Chroniclers of De Soto's Expedition," The Mississippi Historical Society, *Publications,* VII, 379-387; J. A. Robertson's

With regard to the account of Ranjel, De Soto's private secretary, Oviedo wrote as follows:

> And so this *Rodrigo Ranjel*, after all these things had happened, . . . *came to this city of Santo Domingo, in the Island of Española,* and gave an account of all these things in this royal audiencia, [to the most Reverend Licentiate Alonso López de Cerrato, its president, who] asked him and charged him that he should give me in writing an account of everything, in order that, as their Majesties' chronicler of these histories of the Indies, I might gather together and include therein this conquest and discovery in the North. . . .[4]

The words italicized in the above quotation are thus commented upon by Bourne: "This sentence makes it possible to fix the date *before* which Oviedo secured his material, for he left Santo Domingo in August 1546." The words in brackets, which were later deleted by Oviedo and are omitted in Bourne's translation, make it possible to fix the date *after* which Ranjel gave his account of the De Soto expedition to Oviedo, for Alonso López de Cerrato arrived at Santo Domingo, January 1, 1544.[5]

As for the narrative by the Gentleman of Elvas, the printing of it "was finished on the tenth of February, of the year one thousand five hundred and fifty seven in the noble and ever loyal city of Evora,"[6] in Portugal. Hence we may presume that it was given to the printer sometime in 1556, and that it was written sometime after 1544.

According to Ranjel's narrative in Oviedo, the day after leaving Xuala, the expeditionaries,

> wading up to their shins, crossed the river by which they were later to depart in the brigantines which they had made. This river, when it reaches the sea, the chart indicates to be the *Rio del Spiritu Sancto,* which according to the maps of the geographer Alonso de Chaves, empties into a great bay, and the mouth of this river, where the water is salt, is thirty-one degrees north of the equator.[7]

---

translation of the account by the Gentleman of Elvas, *infra,* II, 397-412; *Final Report,* 4-11.

[4] Oviedo, *Historia General y Natural de las Indias,* I, 560; the translation is largely that of Bourne, *Narratives,* II, 48.

[5] Oviedo, I, 158.

[6] Bourne, *Narratives,* I, 223.

[7] *Ibid.,* II, 104.

In his edition of the Ranjel account, Bourne annotated this passage, saying in part:

> The map of Alonso de Chaves, which was constructed in 1536, is no longer extant . . . J. G. Shea (in Winsor, *Narrative and Critical History*, II, 247) assumed that De Soto had this map and studied it; but in the judgment of the present editor the remarks in the text about Chaves' map are by Oviedo, and not derived from Ranjel's diary, and consequently by no means warrant the notion that De Soto and his officers "pored over the cosmography of Alonso de Chaves."

It is quite true that the "remarks in the text about Chaves' map are by Oviedo," and that they do "not warrant the notion" mentioned in the above quotation, for from the description of the map of Chaves by Oviedo, it does not appear that the course of the Rio del Espíritu Santo was inscribed on this map, since it was primarily a sea chart; hence its study by the leaders of the expedition would have been of no avail. De Soto and his officers may not have "pored over the cosmography of Alonso de Chaves," but we know that De Soto spent the last months of 1537 and the first three months of 1538 in Seville;[8] and since his expedition had the official sanction of the Spanish government, his pilots were given maps of the Gulf of Mexico based on the padrón kept in the Casa de Contratación, which had been brought up to date in 1536 by Alonso de Chaves.

In Ranjel's account therefore, the single mention of the Rio del Espíritu Santo is an interpolation by Oviedo. The name does not appear at all in the Elvas narrative with reference to the Mississippi;[9] but in the report of Luis Hernández de Biedma, the *factor* of the expedition, there are four references to a Rio del Espíritu Santo, one of which clearly applies to the Rio Grande of the other two chroniclers, *i.e.*, to the Mississippi.

The first of these four references is in the following passage:

> On the arrival of the brigantines, the Governor directed that they should sail westward to discover a harbor, if one were near, and thence to ascertain, by exploring the coast, if any

---

[8] *Ibid.*, I, 9-11.
[9] The Bahia del Espíritu Santo of those two accounts of the expedition is not the bay in which the river of that name disembogues, but the bay on the west coast of Florida where De Soto landed.

thing could be found inland. Francisco Maldonado . . . had the command. He coasted along the shore, and entered all the coves, creeks and rivers which he discovered, until he arrived at a river having a good entrance and harbor, with an Indian town on the sea board. Some inhabitants approaching to traffic, he took one of them, and directly turned back with him to join us. On this voyage he was absent two months. . . .

After Maldonado got back, the Governor told him that, as we were about to set off in quest of the country which that Indian stated to be on another sea, he must return with the brigantines to Cuba, . . . and if within six months' time he should hear nothing of us, to come with the brigantines, and sail along the shore as far as the *Rio del Expiritusanto,* to which we would resort.[10]

The other two chroniclers of the expedition, Ranjel and the Gentleman of Elvas, also mention the Maldonado episode, but neither refer to the Rio del Espíritu Santo as the place where De Soto expected the ships to meet him six months later, for both designate the rendezvous by two variants of an Indian name, Achuse and Ochus. The text of Ranjel reads as follows:

Maldonado was despatched along the shore with the brigantines to discover a harbor to the west, . . . Maldonado discovered an excellent harbor and brought an Indian from a province adjacent to this coast which was called Achuse, . . . Captain Maldonado was sent to Havana . . . with the instructions and command of the Governor that he should return to the port that he had discovered and to that coast where the Governor expected to arrive.[11]

The same incident is thus narrated by the Gentleman of Elvas:

The Governor directed Francisco Maldonado . . . to sail along the coast to the westward with fifty men, and look for an entrance; proposing to go himself in that direction by land, on a voyage of discovery. . . .

The Captain [returned], bringing with him an Indian from a Province called Ochus, sixty leagues from Apalache, and the news of having found a sheltered port with a good depth of water. The Governor . . . sent Maldonado to Havana for provisions with which to meet him at that port of his discovery, to which he would himself come by land; but should he not reach there that summer, then he directed him

---

[10]Bourne, *Narratives,* II, 8-9.
[11]*Ibid.,* II, 81-82.

to go back to Havana and return there the next season to await him, as he would make it his express object to march in quest of Ochus.[12]

Since two out of three independent witnesses agree as to the name of the meeting place, their statement is to be preferred to that of the third which differs from them; especially since the port of Ochus—Mobile or more probably Pensacola Bay[13]—was known to Biedma.[14]

The second passage in Biedma's report wherein the Rio del Espíritu Santo is mentioned occurs where he speaks of the Xuala region. This part of the country, he says, is hilly; "among these ridges we discovered the source of the great river which we navigated to get out of [Florida] and which we believe to be the *rio de Espiritusanto* . . ."[15] As can be seen, this opinion of Biedma is couched in less categorical terms than Oviedo's assertion in connection with Ranjel's account. In this same passage, the chronicler goes on to say that the expedition went from the Xuala region to Guasuli, an Indian village on the Hiwassee River, and from Guasuli to a town called Chiha [Chiaha], "secluded on an island [Burns Island in the Tennessee River] of this *rio de Espiritusanto,* which all the way from the place of its rise, forms very large islands."

The last time the name occurs is when the Spaniards reached Quizquiz, a town, says Biedma, which "was near the banks of the *rio del Expiritusanto.*"[16] It is quite certain that this town was near the Mississippi, and hence a reader might suppose that the Mississippi was indeed known to the Spaniards as the Rio del Espíritu Santo. Such reasoning, however, is invalid, because neither Ranjel nor Elvas ever call the Mississippi by any other name than Rio Grande, and Biedma himself, who mentions the Mississippi six times after Quizquiz, always refers to it as Rio Grande.

We know that the expedition actually left Florida by way of

---

[12]*Ibid.,* I, 49.
[13]Cf. *Final Report,* 164.
[14]Bourne, *Narratives,* II, 17.
[15]*Ibid.,* II, 15. "Entre estas sierras allamos el nacimiento del rio grande por donde nosotros salimos, e crehemos ser rio de Espiritusanto."
[16]*Ibid.,* II, 25.

the Mississippi. The question is what reason did Biedma have for asserting, in the second of the above passages, that the river he mentions was the headwaters of the Mississippi? This statement could be defended if the expedition had followed the Tennessee to its junction with the Ohio and if they had descended the latter to the Mississippi, for the river mentioned by Biedma is very probably the Little Tennessee. It is quite certain, however, that their route was very different. The army did not follow this tributary to its junction with the Tennessee River, but left it after a few leagues, and went west overland to the Hiwassee. After descending this for a while, they headed west overland again until at last they came upon the Tennessee River above Chattanooga. The Tennessee was followed to Gunthersville, Alabama; at which point the expedition marched due south to the Coosa River. From here on, the general direction of their route is first south along the river to Mabila, situated in Clark County, Alabama; then north-northwest to the Chickasaw towns located in the Pontotoc-Chickasaw counties, Mississippi, and finally due west until they reached the Mississippi between the Arkansas River and the St. Francis River. From all this it is clear that the actual route of the expedition gave Biedma no reason for identifying the headwaters of the Little Tennessee with those of the great river which they descended when they left Florida.

We must now inquire what led Biedma to believe that this great river was the Rio del Espíritu Santo. His report as we know, found "its way into the Spanish archives or offices as early as 1544";[17] and "is the only one of the four [main narratives of the expedition] of which the original manuscript has been preserved."[18] This manuscript, however, is not the original diary or field notes of Biedma, but rather a short report based on his original record, and written after his return to Spain with a view of giving the Spanish government a general, summary account of the expedition. As we shall see, on a map made at the Casa de Contratación to illustrate the De Soto expedition, there

[17] *Relaçam verdadeira . . . feita per hũ fidalgo Deluas,* Evora, 1557, facsimile reproduction, J. A. Robertson, ed., Deland, 1932; the translation, in the second volume, Deland, 1933, is entitled, *True Relation . . . set forth by a Gentleman of Elvas,* II, 406.

[18] *Final Report,* 6.

are quite a few names of Indian towns which are not found in any known narrative. The appearance of these names may be explained by the hypothesis that the geographers of the Casa had at their disposal not only the short report of Biedma, but some fuller account of the Southern United States, such as might have been contained in Biedma's original field notes.

From the wording of the passage under examination, Biedma is clearly expressing his belief at the time of writing that the name of the river is Rio del Espíritu Santo. As we have already noted, in his last six mentions of the Mississippi he calls it Rio Grande, the only name given to the Mississippi by Ranjel and Elvas. The reason for identifying the Mississippi with the Rio del Espíritu Santo at this place is because at the time when he wrote his report he had access to a map of the Gulf of Mexico and wrote the account of his voyage in the light of this map. This would explain why he is the only one of the three chroniclers to speak of the Rio del Espíritu Santo as the meeting place assigned by De Soto to Maldonado; in spite of the fact that he knew of the port of Ochus mentioned by the other two chroniclers. This Rio del Espíritu Santo was certainly not the Mississippi, for there was not then, any more than there is now "a good entrance and harbor" at its mouth. On the map which he had before him, there was no port of Ochus, but in the northwestern corner of the Gulf a large bay was shown as the mouth of the Rio del Espíritu Santo, which was certainly not the Mississippi; and it is this feature that led Biedma to write that De Soto instructed Maldonado on his return from Cuba "to sail along the shore" as far as this prominent landmark on the coast. The map which Biedma had was quite likely a copy or a variant of the Chaves lost chart; for of the three accounts, Biedma's is the only one giving "Bayahonda" [Bahía Honda] as the landing place of the expedition, and this legend first appears in the vicinity of Tampa Bay on the Chaves map of 1536.

It is very likely, then, that Biedma applied the name Rio del Espíritu Santo to the great river because he saw that river so marked on a coastal map of the Gulf. From a similar map, the author of the so-called De Soto map obtained the names of all the rivers emptying into the Gulf on the map which he drew.

The authorship and approximate date of this map was the subject of a recent study.[19] That the author was Alonso de Santa Cruz seems confirmed by the following consideration. On the maps of the other geographers of the Casa, two large rivers, the "Rio Escondido," on the Texas Coast, and the "Rio del Cañaveral or simply "el Cañaveral" east of the Bahía del Espíritu Santo, are always indicated, whereas these two rivers do not appear on any of the known Santa Cruz maps, and are not marked on the De Soto map.

The date of the map is made to depend on the account or accounts utilized by the cartographer. One of these is said to be probably "Ranjel's complete report," that is, Ranjel's diary itself rather than Oviedo's version of it; "for all such reports of recent voyages had to be communicated to the Casa immediately upon the return to Spain of the members of the expedition."[20] This is very true, but the duty of reporting to the Casa was incumbent upon the pilots or the government officials, and Ranjel was simply De Soto's private secretary. "Exactly when Ranjel turned over to the government the report written from his diary of the journey is not known."[21] Actually we have no evidence that Ranjel ever handed a report to the government in Spain. We only know that De Soto's secretary told his story to Cerrato, the president of the Santo Domingo Audiencia, who charged him to write a report and give it to Oviedo. There are admittedly "a number of names on the map which do not appear in any known account," but this does not mean that they are taken from Ranjel's diary. If, as seems probable from our previous considerations, Biedma's report was based on a diary or on field notes, these were communicated to the geographers of the Casa, for Biedma was an official and as such it was his duty to report to Seville.

For the coast line Santa Cruz made use of the Chaves padrón of 1536 with slight modifications. On this as on all early official sea charts only the mouths of the rivers emptying into the Gulf were shown. When he read, however, that the headwaters of the great river were in the foothills west of Xuala, that it formed large islands and flowed past Quizquiz, and when he further noted Biedma's belief that this was the Rio del Espíritu Santo,

---

[19] B. Boston, "The 'De Soto Map'," *Mid-America*, XXIII, 1941, 236-250.
[20] *Ibid.*, 245.
[21] *Ibid.*, 246.

since on the map which he had, as well as on that which Biedma had, a Rio del Espíritu Santo was shown emptying into the Gulf in the northwest corner, Santa Cruz drew the river accordingly, joining the three points and showing large islands in its upper course.

What we said about the course of rivers shown on the maps of the second group applies equally to the course of rivers represented on this map. After all, the geographical department of the Casa was founded with the express purpose of providing pilots with *sea* charts, and the accuracy of the representation of the interior, the position of Indian villages, the course of rivers, the direction of mountain ranges, were of less importance than the location of coastal islands and the accurate delineation of the coast line. We do not mean, of course, that the Sevilian geographers were indifferent with regard to presenting a fairly reliable representation of the American hinterland; we merely mean that this was a secondary consideration with them. It must also be remembered that a chronicler of an inland journey had less need of accuracy than a ship pilot, and was consequently more careless with regard to latitudes, distances and directions.

There is no evidence that the hydrography and the inland nomenclature were inserted in the padrón, and there was no law against geographers embodying in their own maps information derived from narratives of travels in the interior. Whether mapmakers belonged to the Casa de Contratación or not, they did not put their informant under oath nor did they require his sworn testimony to be corroborated by witnesses—two conditions for insertion in the padrón—before accepting and inserting the data of an inland journey in the unofficial regional maps of the interior which they happened to draw. The De Soto map is a case in point. How could a member of the expedition have known, and therefore, how could Santa Cruz have known that the river passing by Cossa, Ayataba, Talissi, Tascalussa, and Tiachi was the Rio de Flores of the old maps based on the padrón? that Alibano, Chicasa, Pafalaya were on the Rio de los Angeles? that the river which passed through Chaguet was the Rio de Montañas? et cetera. To know these details, the Spaniards would have had to descend each of these rivers to the Gulf, identify the mouth of each as being that marked on the sea chart, then reascend the river to its headwaters or proceed along the coast to the next river and explore it upstream.

Even though, as I believe, Santa Cruz had access to the diary or field notes of Biedma and made use thereof to draw his map, still the course of the rivers on the De Soto map can hardly be said to resemble the hydrography of the Southern United States. For example, what is the giant river which forks into two branches 120 miles above the Gulf and whose two mouths, marked Rio de Pescadores and Rio del Oro respectively, are one hundred miles apart on the coast of Texas? What is the other nameless river with one mouth in the Gulf and another in the Atlantic?

The Rio del Espíritu Santo on this map has multiple headwaters, two in the east and two in the north. The easternmost rises in the foothills of a mountain range at the top of the map; the longitude of this source is due north of Bahía Honda, our Tampa Bay, while the longitude of the south branch of this eastern source is due north of Apalachee Bay. These two branches run westward, north of the mountain range in which they originate, as far as Pacoa [Arkansas], where five branches of the river form three islands. From this point it flows due south to the "baya del espiritu santo." Halfway down its way to the Gulf it receives a tributary coming from the north-northwest, which in turn is represented as formed by the confluence of two rivers with their headwaters in a northern mountain range. The bay into which it empties communicates with the sea by means of a large body of water called "mar pequeña," located as on the other maps in the northwest corner of the Gulf. This representation of the Rio del Espíritu Santo can hardly be said to correspond with the known course of the Mississippi or with the known course of any river east of it. The rivers which Biedma saw in the east are indeed subtributaries of the Mississippi, but neither he nor any other member of the expedition followed them to the main stream, for, as we saw, they left the Tennessee at Guntersville, Alabama.

This De Soto map is a draft which Santa Cruz did not finish, or if he did, the finished copy has not come to light. It does not seem that Juan López de Velasco, his successor as *cosmographo mayor,* is referring to the De Soto map in his description of

America, but rather to a sketchier, although later[22] map of the Southern United States which he found among the papers of Santa Cruz. This deduction is based on his description of the "Gulf of New Spain or Florida." The survivors of the De Soto expedition, he wrote, found the land of Florida fertile; they found grapes, nuts, and other fruits similar to those of Spain; there were pelts and many indications of pearls but "nobody went back to these provinces, and thus this region is not known, nor are the villages thereof, except what is represented on the sketches of Santa Cruz."[23]

The specific text on which we base our deduction that López de Velasco had before him a map different from the so-called De Soto map, occurs in his detailed description of the Gulf coast line "according to the maps of Santa Cruz."

> *Rio y Bahía del Espíritu Santo:* it comes down from latitude 37°, running east-west along this parallel from the meridian of Florida until it is due north of that bay [Bahía del Espíritu Santo], into which the river enters. The bay lies in latitude 31°, thirty leagues or more to the west of the Rio Bajo; it is a great bay into which four rivers empty; at the entrance, close to the east side, there is a small island.[24]

Only one river empties into the Bahía del Espíritu Santo on the De Soto map; no island is indicated near the opening of this bay; and no latitudes are marked on this map.

Between the date when the De Soto map was made and the publication of a simplified version of it in 1584, the Luna expedition took place. The editor of the Luna papers identifies the Rio del Espíritu Santo spoken of therein as the Mississippi,[25] but the passages in which the name occurs hardly justify this identification. With one exception, all occurrences of the name Rio del Espíritu Santo are found in the letters of Luís de Velasco, the viceroy of New Spain. Since at this date there were certainly

---

[22]Cf. *supra,* p. 5.

[23]"Y asi no se tiene noticia de lo que son, ni de las poblaciones que en ella van descritas, más de la que se halla pintada en la descripciones de Santa Cruz"; Lopéz de Velasco, *Geographia y Descripción Universal de las Indias,* 180.

[24]*Ibid.,* 181.

[25]See the index to *The Luna Papers,* H. I. Priestley, ed., 2 volumes, Deland, Fla., 1938.

maps of America in Mexico City, and since Velasco knew approximately where the expedition had landed on the north coast of the Gulf, he could see by consulting any map that the Rio del Espíritu Santo was west of Pensacola, and consequently it was only natural for him to designate this river as a landmark of western explorations.[26] From his use of the name in these passages therefore, we have no right to conclude that "the Mississippi is clearly meant."[27]

In one of these letters, Velasco said that horses were to be sent "provided a route be found over which it is possible to go until they strike the Rio del Espíritu Santo at a place where they can cross it."[28] In the editor's index "to cross" is here interpreted as meaning "to ford." This interpretation is supported by what we read in Father Dávila Padilla's book[29] where he, too, speaks of the Rio del Espíritu Santo as fordable. This latter passage will be given in its context so that the reader may judge whether the location or description of the Rio del Espíritu Santo by the Dominican chronicler corresponds with the Mississippi.

The survivors of the De Soto expedition had given out that the Province of Cossa, the chief village of which bore the same name, was the most fertile of all Florida. In the hope of obtaining badly needed provisions a detachment of soldiers was sent thither from Achuse. After fifty days' march, the party reached Olibahali (Ulibahali in Ranjel, Ullibahali in Elvas), a town situated on the Tallapoosa River in Elmore County, Alabama. Sometime in July they reached Cossa or Coça, located on the Coosa River, in Talladega County, Alabama. The Indians of this village asked assistance of the Spaniards for a war expedition against the Napochies, identified by Shea as being "in all probability the Natchez,"[30] while Gatschet thinks they were the Napissa,[31]

---

[26]*Ibid.,* I, 73, 85, 95.
[27]*Ibid.,* introduction, xxxii.
[28]*Ibid.,* I, 65.
[29]A. Dávila Padilla, *Historia de la Fvndacion y discurso de la Prouincia de Santiago de Mexico, de la Orden de Predicadores,* . . . Madrid, 1596.
[30]Winsor, *Narrative and Critical History,* II, 258.
[31]A. S. Gatschet, *A Migration Legend of the Creek Indians,* 2 volumes, I, Philadelphia, 1884, II, St. Louis, 1888, I, 99, 112, 190.

"probably a Muskhogean people more nearly affiliated to the Choctaw."[32]

The Spaniards agreed to help the Coças [Creeks], and the expedition composed of three hundred of these Indians and fifty Spaniards left the village heading westward. After a few days' march they reached the first Napochies village, which they found empty. The Coças explained that the Napochies, knowing that the Spaniards had joined force with them and distrusting the security of the woods (montes), had gone off on the great water to hide themselves (se fueron a escôder en la grâ agua).

When the Spaniards heard the name of "great water," they thought that it must be the sea; but it was nothing else than a great river (rio grande) which we call the [Rio] del Espíritu Santo, which rises in the great mountains of that land of Florida. It is very deep, and two musket-shots wide. At a certain place known to the Indians, the river is much wider, losing its depth, [so that] it can be forded (y podia vadearse). The Napochies of the first village had crossed it, and those of the other [Napochies] village on the bank of this river, when hearing the news [of the coming of the Coças and the Spaniards], also abandoned it, crossing the waters of Oquechiton, for thus the Indians call this river, that is to say, in our language, the great water (la grãde agua).

As Father Dávila Padilla has just said, Coças and Spaniards found the second Napochies village also empty when they arrived.

The people of both villages were on the other bank of the river fully confident that the Spaniards had no means of crossing it; they were loud in their vociferations and jibbed at the Coças. Their joy did not last long, for the Coças, who knew the country well, found the ford of the river and entered it, the water reaching the breasts of those on foot and the saddles of those on horseback . . . When our soldiers reached the middle of the river, one of them discharged his gun, loaded with two bullets, and brought down a Napochies who was on the other bank of the river.[33]

The terrified Napochies took to flight pursued by the Coças;

---

[32]F. W. Hodge, ed., *Handbook of American Indians North of Mexico*, 2 parts, Washington, D. C., 1912⁴, *s. v.* Napochies.
[33]*Historia de la Fvndacion*, 262.

the fugitives were finally subdued and accepted the Coças peace terms.

> This done the whole expedition returned to the first Napochies village, where a garrison of Spaniards and Coças had been left. Since this village was convenient, our men rested there three days, until it seemed time to return to Coça, where 150 Spanish soldiers were waiting. The journey was short and they soon arrived there."[34]

According to Shea the Napochies were "a nation near Ochechiton [Oquechiton], the Espíritu Santo, or the Mississippi,"[35] and, as we saw, he thought the Napochies were the Natchez. In his summary of Father Dávila Padilla's narrative, Lowery says: "The Napochies were pursued to and even beyond the Oquechiton —the Mississippi River"; adding in a note the meaning of the word given by the Dominican, namely, " 'The Great Water,' the Espíritu Santo."[36] Father Steck combined what Shea and Lowery wrote, saying: "At the same time, a detachment of soldiers joining the friendly Coça Indians in their war with the Napochies (Natchez) actually reached and most probably also crossed the Mississippi River."[37] If Oquechiton is the Mississippi the Spaniards crossed it; the text of Father Dávila Padilla is quite clear in this respect: after an Indian had been shot from midstream, "all the Napochies took to flight, and the Coças crossed the river, keeping at fire-arm range."[38]

Whatever the meaning of the word "Oquechiton" may be,[39] this river was not the Rio del Espíritu Santo, which is located on

---

[34]*Ibid.*, 263.

[35]Winsor, *Narrative and Critical History*, II, 258.

[36]Lowery, *The Spanish Settlements*, 367.

[37]F. B. Steck, *The Jolliet-Marquette Expedition 1673*, Quincy, Illinois, 1928, 203.

[38]*História de la Fvndacion*, 263.

[39]"Ochechiton, like Mississippi, means great river—from *okhina,* river; *chito,* great; [Cyrus] (Byington's [An English and] Choctaw Definer, [New York, 1852], pp. 79, 97)," Shea, in Winsor, *Narrative and Critical History of America,* II, 258, note 7. Rather "great water," as Father Dávila Padilla has it, from *oka,* water, and *chito,* great, large; cf. *A Dictionary of the Choctaw Language by Cyrus Byington,* edited by J. R. Swanton and H. S. Halbert, Smithsonian Institution, Bureau of American Ethnology, *Bulletin 46,* Washington, D. C., 1915, 107, 291.

the maps much farther west than the actual location of the Mississippi, nor was it the Mississippi, whether one considers its location or its description.

The site of Coça, the starting point of the expedition against the Napochies, is known with a fair degree of accuracy. It was situated on the Coosa River in Talladega County, Alabama;[40] that is, more than 300 miles in a straight line east of the Mississippi. In one day's march, according to Father Dávila Padilla, the party covered about eight leagues,[41] and it took two or three days to reach the first Napochies village from Coça. Elsewhere, the chronicler again speaks of the distance between Coça and the first Napochies village: "Era el camino breve y llegaron presto." The second Napochies village was close by the first and near Oquechiton, so that the whole distance between Coça and the "great water" was at the most thirty leagues, or about one third of the distance between Coça and the Mississippi. Again, it took the Spaniards two months to cover the distance from Achuse to the Coça village, that is, it took them twenty times longer to travel over a distance shorter by fifty miles than that between Coça and the Mississippi.

With regard to the description of the river, the Mississippi is not fordable anywhere near the supposed point of crossing. Father Dávila Padilla specifies the depth of the river as less than five feet; and he speaks of a soldier firing his gun while standing in the middle of the stream; but the Mississippi in that part of its course is wider than twice the range of a sixteenth-century musket. Whatever the identity of the Oquechiton may be, the evidence in Father Dávila Padilla's chronicle does not justify its identification with the Rio del Espíritu Santo which appears on contemporary maps or with the Mississippi.

A Rio del Espíritu Santo is also mentioned in the narrative of the Coronado expedition by Castañeda, whose knowledge of the geography of North America was very indefinite. It is believed, he wrote, that the Tiguex [our Rio Grande], "flows into the mighty (poderoso) Rio del Espíritu Santo which the men with Don Hernando de Soto discovered in Florida."[42] From this

---
[40] *Final Report,* 206.
[41] *Historia de la Fvndacion,* 254.
[42] Winship, *The Coronado Expedition,* 510.

and from other passages of his narrative to be quoted presently, it seems that Castañeda had, in Culiacán, where the account of the Coronado expedition is believed to have been written, some freak map of the Western hemisphere, into which he tried to fit what he learned from men who took part in De Soto's expedition. Whatever these men may have told him, we saw that neither Ranjel nor Elvas speak of the great river discovered by De Soto as the Rio del Espíritu Santo, and, we may add, Garcilaso de la Vega's informants, who were also members of the De Soto expedition, call the great river Chucagua.

Castañeda speaks twice of the Rio del Espíritu Santo, and from his description some have recognized not only the Mississippi but the Missouri as well.[43] The following quotation, the first of the two mentions of the Rio del Espíritu Santo, will show whether Castañeda's description enables one to identify this river as the Mississippi.

> The great Rio del Espíritu Santo, which Don Fernando de Soto discovered in the country of Florida flows through this country [Quivira]. It passes through a province called Arache, according to the reliable accounts which were obtained here [Culiacán]. The sources were not visited, because, according to what they said, it comes from a very distant country in the mountains of the South Sea [Pacific Ocean], from the part that sheds its waters onto the plains. It flows across all the level country and breaks through the mountains of the North Sea [Atlantic Ocean], and comes out where the people with Don Fernando de Soto navigated it. This is more than 300 leagues from where it enters the sea. On account of this, and also because it has large tributaries, it is so mighty when it enters the sea that they lost sight of the land before the water ceased to be fresh.[44]

The "mountains of the North Sea" is the mountain range which geographers imagined ran east-west along the thirty-fifth parallel. On no map is there a river shown breaking through this range, for Florida is always represented as a closed amphitheater, and the Mississippi does not break through mountains at the point where De Soto and his men saw it.

---

[43]*Ibid.*, 529. "The Missouri-Mississippi" is Hodge's slightly different identification in *Spanish Explorers in the Southern United States*, 365, note 2.

[44]Winship, *The Coronado Expedition*, 529.

The next time the Rio del Espíritu Santo is mentioned in the narrative of the Coronado expedition, Castañeda speaks of it as follows:

> It is, I think, already understood that the Portuguese, Campo, was the soldier who escaped when Fray Juan de Padilla was killed at Quivira, and that he finally reached New Spain from Pánuco, having traveled across the plains country until he came to cross the North Sea mountain chain, keeping the country that Don Hernando de Soto discovered all the time on his left hand, since he did not see the Rio del Espíritu Santo at all. After he crossed the North Sea mountains he found he was in Pánuco, so that if he had not tried to go to the North Sea, he could have come out in the neighborhood of the borderland, or the country of the Sacatecas, of which we now have some knowledge.[45]

Castañeda introduces the above quotation with the words: 'I very much wish that I possessed some knowledge of cosmography and geography so as to render what I wish to say intelligible." This is also our wish. For when we consider the actual geography of the country described, and when we compare what he is saying here with what he says in the preceding quotation, his geographical digression is quite unintelligible.

As was said above, a version of the De Soto map was published in 1584.[46] It is this simplified variant which was to influence mapmakers for the next forty years, for the map of Santa Cruz did not come to light until the nineteenth century. The author of the map of Florida which was published in 1584 is Hierónymo,[47]

---

[45] *Ibid.*, 544.

[46] It was published simultaneously in A. Ortelius, *Theatrum Orbis Terrarum,* Antwerp, 1584, and in the *Additamentum III Theatri Orbis Terrarum,* Antwerp, 1584. The map of Florida in the 1579 edition of the *Theatrum* in the Library of Congress is one of the twenty maps bound in at a later date with the 93 maps which made up the original 1579 edition; cf. P. L. Phillips, *A List of Geographical Atlases in the Library of Congress,* Washington, D. C., 1909, n. 386, and the note of Ortelius on the title page of the *Additamentum III.—Infra,* pl. 5. (III)

[47] On Hierónymo or Gerónimo de Chaves, cf. Fernández de Navarrette, *Biblioteca Maritima Española,* 2 volumes, Madrid, 1851, I, 563-565; summarized by L. Bagrow, "A. Ortelli Catalogus Cartographorum," *Petermanns Mitteilungen,* XLIII, 1929, Ergänzungsheft, n. 199, 1928, 55-56; and M. Jiménez de la Espada, *Relaciones Geográficas de Indias,* 4 volumes, Madrid, 1881-1897, III, v-vii.

the son of the royal cosmographer Alonso de Chaves, but it is impossible to determine in what year it was made. All that can be said is that it is prior to 1574, the year of Hierónymo de Chaves' death. It is not likely that the manuscript was in Ortelius' possession much before 1584, because this map is not found in the Second Supplement published in 1580[48] in which there are three special maps of North America not found in the earlier editions of the *Theatrum*. There is no doubt, however, that this published map of Florida is a simplified version of a variant of a map similar to that of Santa Cruz illustrating the discoveries of De Soto in the Southern United States.

A comparison between the nomenclature and the hydrography on the two maps yields interesting data. On the De Soto map there are three times—sixty-six—as many place-names as on the Chaves map of Florida—twenty-two. The nomenclature coincides on both maps, except that Chaves has three place-names that are not on the manuscript map, although they are mentioned by the chroniclers: Catilachegue, Achusi, and Chillano. The first is a variant of Cofitacheque (Ranjel), Cofitachique (Biedma), Cutifachiqui (Elvas). The fantastic spelling of Indian names by Spanish writers would be reason enough for the difference between Catilacheque and the three variants. In the present case, matters were made still worse by the fact that the draughtsman and the engraver, both Dutch, copied an outlandish name from

---

[48] The name of the Spanish cartographer is listed thus in the catalogue of authors in the first edition of the *Theatrum:* "Hieronymus Chiauez, Americam descripsit quae nondum in lucem prodiit." In the Latin edition of 1573, Ortelius added to this notice: "Idem Conuentum Hispalensem." These words are repeated in the other editions until that of 1579, where we read: "Hieronymus Chiauez, Americam descripsit, quae nondum in lucem prodiit. Idem Conuentum Hispalensem, quem iam in hoc Theatro in lucem damus." This is a map of Seville and the adjacent territory. This notice was not changed in the subsequent editions of the *Theatrum;* Ortelius does not mention the fact that he included Chaves' map of Florida in the edition of 1584. The words "Americam descripsit" very probably refer to a map. Fernandez de Navarrete, *Biblioteca Marítima*, I, 565, says that among the papers of Chaves are "1) Mapa del territorio de Sevilla, que puso A. Ortelius en su Teatro," this is the Conuentus Hispalensis; "2) Otro de las Indias Occidentales, que estaba inédito en la libreria del conde de Villaumbrosa; (y hay copia en el Depósito hidrografico en Madrid, tom. 17 de Mss)." I have not seen this map.

a map made by a Spaniard. Achusi, the second name omitted by Santa Cruz but inserted by Chaves in his map, is clearly Ranjel's Achuse, Elvas' Ochus. As we have seen, Biedma does not give this name of the port where Maldonado went, but later in the narrative, he speaks of the Bay of Chuse, a variant of Achuse. The third place-name occurs only in the narrative of the Gentleman of Elvas. Whether it was also in Ranjel's account cannot be known, for his report in Oviedo breaks off at the beginning of November 1541,[49] and Chillano was discovered more than a year later, at the end of November 1542.

Before drawing conclusions from the differences in nomenclature on the two maps, it is necessary to recall certain previously indicated facts or deductions. First, the Biedma account is a short report seemingly based on a diary or field notes accessible to the geographers of the Casa, who made use of the information contained therein when drawing their maps. Secondly, the fact that some place-names are found in the narratives and not on the Santa Cruz map is due to the state of the map, which is obviously an unfinished draft. Thus the place-names in the east are much fewer than in the west, and the map has the sites of sixteen villages marked but left nameless. Finally, even if we had not the testimony of López de Velasco that there was at least another map of the Southern United States by Santa Cruz, it is quite unlikely that only one map illustrating the De Soto expedition was made in Seville, and still more unlikely that this unique map would be precisely the one which has come down to us.

The reports handed to the Casa de Contratación were available to all the members of the geographical department; all had access to the padrón; all could easily know the maps made by their colleagues. Now Hierónymo de Chaves was one of the cosmographers of the Casa. As a basis for his map of Florida, he made use for the coast line of a map slightly different from that of Santa Cruz and more closely related to the map of 1536 made by his father, filling the interior with place-names first made known by the survivors of the De Soto expedition. In the present

---

[49]Chapter XXVIII in Oviedo is incomplete. Chapter XXIX began with the narrative of the death of De Soto, which occurred nearly seven months after the last date mentioned in Chapter XXVIII.

case, according to our hypothesis, he used the names which he read in Biedma's diary or field notes. Considering that the Santa Cruz map is an unfinished draft and that there are on the printed map three place-names which are not on the extant manuscript map of the De Soto expedition, we conclude that the nomenclature of the lost Chaves original was very probably much fuller. It is very unlikely that Chaves himself selected from the many towns and villages those which appear on the printed map. The selection was more probably made by the draughtsman or by the engraver; for the exigencies of space determined the number of place-names on this map, as in the case of the simplified version of Mercator's planisphere of 1569 in Ortelius' atlas of 1570.

Even more pertinent to our inquiry than this comparison of the inland nomenclature on the two maps is the hydrography of the Southern United States as represented on both of them. The Chaves map of the Gulf Coast, as we just noted, was different from that used by Santa Cruz, and attention has already been called to the fact that the course of the rivers on the De Soto map was conjectural and that their names were determined by the names inscribed near the mouth on the coastal map used as a basis. Three examples taken from the map of Santa Cruz and from that of Chaves may serve to illustrate these points.

Since no authentic Santa Cruz map contains the Rio Escondido, which empties into the Gulf on the Texas Coast, the course of a river of that name is not shown on the De Soto map. On the other hand, maps made by other cartographers of the Casa show the mouth of the Rio Escondido in the Texas region; consequently Chaves, who used a coastal map different from that used by Santa Cruz, drew the course of this river along latitude 27°30′, that is, just south of Corpus Christi Pass. Where this Rio Escondido had its headwaters cannot be known because the border of the map cuts it short 170 miles from the sea.

The second example is more striking. On the Chaves map, east of the Rio del Espíritu Santo, there is a giant river, Rio de Cañaveral, running parallel to it for nearly 500 miles, from latitude 37°30′, *i.e.*, half a degree north of the mouth of the Ohio, where it branches off a large river coming from the east. The latter, at this point, flows to the southwest, emptying into the Rio del Espíritu Santo at latitude 35°. As in the case of the Rio

Escondido, the Rio de Cañaveral never appears on the Santa Cruz maps. On the De Soto map, next to the Rio del Espíritu Santo is a short Rio Baxo rising in the foothills of a mountain near Quigualta. Now there is no Rio Baxo on the Chaves map, and there is no Rio de Cañaveral on the Santa Cruz map. On the latter, the next river east of the Rio Baxo is legended Rio de los Angeles, but there is no trace of such a river on the Chaves map.

The third example, showing how unreliable is the inland hydrography, is that of a large river which, on the De Soto map, divides itself into two branches, one emptying into Apalachee Bay, the other into the Atlantic at latitude 31°, both nameless. On the Chaves map, there is also such a river, but the mouth of each branch is on the Atlantic Coast, and the southern branch is legended Rio Seco, a name inscribed near an indentation of the coast on earlier maps.

These discrepancies are here pointed out not to belittle the cartographical work of the geographers of the Casa, but because they show that the inland hydrography which appears on their maps is conjectural, depending largely on the coastal map used as a basis, and therefore all attempts to identify rivers from their length, the direction of their course, their position, or their name on these maps must necessarily be futile.

With regard to ascertaining the modern equivalent of the position of certain landmarks on these two maps the following is to be said. Neither latitudes nor longitudes are marked on the De Soto map, and no scale is given. We can, however, roughly calculate the positions by using as a scale the Floridan peninsula, the length of which on all the early maps is five degrees of latitude or about 350 miles. According to this scale, the position of the mouth of the Rio del Espíritu Santo is nowhere near that of the mouth of the Mississippi, for the opening of the Mar pequeña —the mouth of the Rio del Espíritu Santo—would be located on a modern map, near Austin, Texas. If we were to convert this distance into degrees, taking as a basis the meridian that grazes the east coast of Florida, the mouth of the Rio del Espíritu Santo would be located on a modern map in the vicinity of Galveston Bay; and if we take into consideration the length of a degree of longitude along the thirtieth parallel, the mouth of the "Mississippi" would again have to be located in the vicinity of

Austin, Texas. By the same method, Neguateix, the westernmost town, should be placed in Otero County, New Mexico.

On the Chaves map, the longitudes and latitudes are marked. The opening of the Mar pequeña is thirteen degrees west of the basic eastern Floridan meridian, or in the vicinity of Cameron, Louisiana. This more easternly position, at variance with that found by previous computations, is due to the fact that on this map one degree of longitude is equal to one degree of latitude. If the mapmaker had observed the correct relation between one degree of longitude and one degree of latitude along the thirtieth parallel, the position of the mouth of the Rio del Espíritu Santo would have been, as on the majority of maps, near Galveston Bay. By the first of these two methods of conversion the northernmost point of the Rio del Cañaveral would be located, on a modern map, in Laclede County, Missouri; by the second method, in Wilson County, southeastern Kansas. Naguater, Neguateix of the De Soto map, would have to be placed either in Graham County, Kansas, or in Park County, southwest of Denver, Colorado, according to which method of conversion was used.

The importance of the Chaves map comes from the great influence it had on subsequent mapmakers, who for forty years after its publication widely imitated or copied outright this new geographical conception of the Southern United States. As such it is a landmark of the cartographical history of the country north of the Gulf of Mexico, and is the type of the third group of maps studied in this essay. Credit for this new geographical conception has thus far been given to Cornelius Wytfliet. The latter's map of Florida, however, did not appear until thirteen years after the publication of the Chaves map by Ortelius, and for all that pertains to the hydrography and nomenclature of the interior it is simply a slavish imitation of the earlier one by the Spanish cartographer. We may note here that on other copies than Wytfliet's the number of place-names along the Gulf Coast and in the interior is greater than on the model because the draughtsman or the engraver wished to "complete" the Chaves coastal or inland nomenclature with names from other published maps.

When we say that the Chaves map was widely imitated by mapmakers for forty years after its publication in 1584, we do not mean that this geographical conception was thereafter discarded, for, as can be seen from the dates of the maps listed below, the Chaves model was still used as late as 1670. In 1625, however, a new type of map of the Southern United States appeared. Just as maps of the first group were still made long after a new type had appeared in print, so maps belonging to the second group are found after the publication of the Chaves map, and maps of the third group, after De Laet's map of Florida of 1625. The selection of a particular type often depended on the fancy of the mapmaker, more often on those maps or atlases he happened to have at hand.

As in the case of the maps of the second group those based on Chaves' may be divided into two classes: the first class comprising maps which reproduce the model with slight alterations, the second including those on which the Chaves' hydrography and nomenclature had to be simplified because of exigencies of space. Since many mapmakers superimposed the Chaves Florida map on Mercator's planisphere of 1569, the nomenclature of this latter is generally found on maps of this group, except some regional maps of the Southern United States.

The map of the Western hemisphere by Mazza belongs to the first class of the third group.[50] Mazza superimposed the inland hydrography and the nomenclature of Chaves' Florida map on the map of the Americas in Ortelius' atlas of 1570.[51] The Southern United States on Plancius' World map of 1592,[52] and the map of North America by Cornelius De Jode published the following year,[53] are nothing else than the Chaves map of Florida

---

[50]Americae et Proximar[um] Regionum Orae Descriptio, in *Remarkable Maps*, I, 12; in *Atlas brésilien*, no. 29. It is dated "ca. 1583" in the first, and 1584, in the second of these two compilations. It is certain, however, that the map is posterior to the *Theatrum* of 1584.

[51]Americae sive Novi Orbis Nova Descriptio, in *Theatrum Orbis Terrarum*, Antwerp, 1570. Bibliography in Lowery, *Descriptive List of Maps of the Spanish Posessions*, 74-76.

[52]Full size reproduction in F. C. Wieder, *Monumenta Cartographica*, 5 volumes, The Hague, 1925-1933, II. Cf. E. Heawood, *The Map of the World on Mercator's Projection by Jodocus Hondius Amsterdam 1608*, London, 1927, 2. Wieder's comments on the geography of New France and Florida as represented on this map, text, p. 31 b, are not very helpful.

[53]Americae Pars Borealis, Florida, Baccalaos, Canada, Corterealis. A Cornelio de Judaeis in luce edita, in *Specvlvm Orbis Terràrvm*, [Antwerp],

superimposed on the southeast section of Mercator's planisphere of 1569. The reason for the changes in coastal nomenclature has been given above; the change of position of towns in the interior is arbitrary and should be attributed to the draughtsman or to the engraver. As we have already noted, the Wytfliet Florida map of 1597 is a copy of that of Chaves,[54] except for the omission of one Indian village, and Matal's map of Florida[55] is a copy of that of Wytfliet.

The following geographers, mapmakers, engravers, or map publishers transposed the hydrography and the nomenclature of Chaves' map of Florida on their own representations of the world, of the Western hemisphere or of North America: M. Tatton on his map of New Spain;[56] William Blaeu on the World map of 1605;[57] and on the map of America in the atlas of 1663;[58] Jodocus Hondius on the world maps of 1608 and 1611,[59] and on the maps of the Americas in the various editions of the larger Merca-

---

1593. Cf. F. Van Ortroy, *L'oeuvre cartographique de Gérard et de Corneille de Jode,* Ghent, 1914, 33.

[54]Florida et Apalche, in *Descriptionis Ptolemaicae Augmentum,* Louvain, 1597, between pp. 176-177. There is a good reproduction of the region north of the Tropic of Cancer in Winsor, *Narrative and Critical History,* II, 281.

[55]Florida et Apalche, in *Speculum Orbis Terrae,* Cologne, 1600-1602.

[56]Noua et rece[ns] Terra[r]um et regnorum . . . uná cum exacta absolutaq[ue] orarum sinus Mexicani . . . delineatio á M. Tattonus . . . edita. The name of the engraver, Benjamin Wright, and the date, 1600, are in the upper right hand corner cartouche. See Lowery, *Descriptive List of Maps,* 118-119. The map is reproduced in I. B. Richman, *California under Spain and Mexico, 1535-1847,* Boston and New York, 1911, 370, map VI.

[57]Nova Universi Terrarum Orbis Mappa . . . duobus planispheris graphicè depicta à Guliel. Janssonio Alcmar [Blaeu], facsimile reproduction published by E. L. Stevenson, under the auspices of the Hispanic Society of America, New York, 1914. Cf. also John Blaeu's Nova Totius Terrarum Orbis Tabula, in Wieder, *Monumenta Cartographica,* III.

[58]*Le Grand atlas ov Cosmographie Blaviane,* vol. XII, Amsterdam, 1663.

[59]Nova et Exacta Totius Orbis Terrarum Descriptio Geographica et Hydrographica. Auct. I. Hondio, (1608), reproduced in E. Heawood, *The Map of the World . . . by Jodocus Hondius,* London, 1927. Novissima ac Exactissima Totius Orbis Terrarum Descriptio Magna . . . Auct. J. Hondio, (1611), reproduced in E. L. Stevenson and J. Fischer, eds., *Map of the World by Jodocus Hondius,* New York, 1907.

tor atlas;[60] on the maps of Van den Keere,[61] Merian,[62] Hoeius,[63] Heylin,[64] Ogilby,[65] et cetera.

The following maps belong to the second class of the third group: Plancius' globe of 1594;[66] De Bry's Western hemisphere;[67] Hondius' planisphere known as "The Christian Knight" map;[68] the Plancius-Vrient mappamundi of 1599;[69] the Western hemisphere in the various editions of the smaller Mercator-Hondius atlas;[70] the Goos gores of 1621;[71] Danckert's mappamundi of 1640,[72] et cetera.

It is clear that none of this cartographical evidence tends to show that the Rio del Espíritu Santo was the Mississippi, for all these mapmakers merely copied Chaves' map on which, as we have seen, the Rio del Espíritu Santo certainly does not represent

---

[60]America, in *Gerardi Mercatoris Atlas sive Cosmographicae Meditationes de Fabrica Mvndi et Fabricati Figvra*, Amsterdam, 1611⁴.

[61]Americae Nova Descriptio, in *Atlas brésilien*, no. 56.

[62]America noviter delineata. M. Merian fecit, cf. Lowery, *Descriptive List of Maps of the Spanish Possessions*, 142.

[63]Nova Orbis Terrarum Geographica ac Hydrographica Descriptio, Ex optimis . . . Tabulis desumpta à Franciscus Hoeius, reproduced from Allardt's edition in *Remarkable Maps*, I, 7-8.

[64]Americae nova descriptio, in P. Heylin, *Cosmographie, The Fourth Book. Part II . . . containing the Chorography and History of America*, London, 1668. The map is dated 1663.

[65]Novissima et Accuratissima Totius Americae Descriptio per Johanem Ogiluium Cosmographum Regium, in *America: being the latest and most accurate Description of the Nevv VVorld*, London, 1671.

[66]Orbis Terrarum Typus de Integro Multis in Locis emendatus. Auctore Petro Plancio, in *Atlas brésilien*, no. 37, and in R. Hakluyt's *Principal navigations*, London, IX, 1904, 474.—*Infra*, pl. 6. (IIIa)

[67]America sive novvs orbis respectu Europaeorum Inferior Globi Terrestris Pars, in Theod. De Bry, *Americae Pars VI*, Frankfurt, 1596. Reproduced in *Atlas brésilien*, no. 40; the section north of the Equator in Winship, *The Coronado Expedition*, facing p. 393.

[68]Typus Totius Orbis Terrarum . . . reproduced, reduced size, in *Atlas brésilien*, no. 47.

[69]Orbis Terrae Compendiosa Descriptio . . . , in *Atlas brésilien*, no. 48.

[70]Americae Descrip[tio], in *Atlas Minor Gerardi Mercatoris á I. Hondio plurimis aeneis tabulis auctus atque illustratus*, Amsterdam, 1607.

[71]In *Remarkable Maps*, I, 9. The cartouche which was to contain the title is blank.

[72]Nova Totius Terrarum Orbis Tabula Auctore D. D., in *Atlas brésilien*, no. 74.

the Mississippi. Of course the name of the river was "kept alive" by these geographers but this fact is of no consequence to our question. The Mercators, the Hondiuses, the Blaeus of the sixteenth and seventeenth centuries compiled their maps from those already on the market, and did not "recognize" the Rio del Espíritu Santo as the greatest of all the rivers emptying into the Gulf of Mexico; for, as their maps clearly show, they could not decide whether the eastern headwaters of the network of the Floridan rivers were those of the Rio del Espíritu Santo or those of the Rio de Cañaveral.

VI

THE FOURTH GROUP OF MAPS— THE SANSON-JAILLOT INTERPRETATION

As was noted above, the Chaves map of Florida, or a simplification of it was very frequently reproduced by European mapmakers from 1584 to 1625, the year when Johan De Laet published a very different map of the Southern United States in his *Nieuwe Wereldt*.[1]

De Laet was an industrious compiler, whose intention was "to give his fellow-citizens as perfect a description of the New World as circumstances would allow."[2] He made use of what he considered to be the best works then in print and gave references to his sources. Thus the chapter on Florida in his book is based on those of López de Gómara, Wytfliet, Herrera,[3] and Huygen,[4]

---

[1] There are four editions of this work, two in Dutch, one in Latin, and one in French. I have been unable to consult the first edition, *Nieuwe Wereldt Ofte Beschrijvinghe van West-Indien*, . . . Leyden, 1625. "This invaluable work was much improved in the subsequent editions and translations by the author, but the maps were unchanged," J. Sabin, *A Dictionary of Books Relating to America*, X, 1878, 15. The same plates were used for the last three issues. According to Winsor, *Narrative and Critical History of America*, IV, 417, the difference between the first and the second editions consists in a slightly changed title. This second edition was published at Leyden, 1630. Three years later, also at Leyden, the Latin edition appeared, *Novus Orbis, seu Descriptionis Indiae Occidentalis Libri XVIII*; and in 1640, in the same town, the French edition, *L'histoire dv Nouveau Monde ou Description des Indes Occidentales Contenant dix-huict Liures* . . . , was published.—*Infra*, pl. 7. (IV)

[2] Winsor, *Narrative and Critical History*, IV, 417. Cf. also J. F. Jameson, *Narratives of New Netherland 1609-1664*, New York, 1909, 31-35.

[3] A. Herrera y Tordesillas, *Historia General de los hechos de los Castellanos en las Islas i Tierra firme del Mar Oceano . . . En quatro Decadas desde el Ano de 1492 hasta el de 1531*, 2 volumes, Madrid, 1601. Bound in the second volume is the *Descripcion de las Indias ocidentales*, with a special title page, also dated Madrid, 1601. Four more decades, in two volumes, were published at Madrid in 1615. The De Soto expedition is narrated in the sixth and seventh decades.

[4] In the bibliography of the 1630 edition, De Laet lists *Ian Huygen van Linschoten Beschrijvinghe van America*, and in the edition of 1640, he gives *Briefue description de l'Amerique par Hugues de Linschot*. I have not seen the Dutch edition of Huygen, I used the *Description de l'Amerique*

but for all that pertains to the subject of our investigation, his main source is Herrera, whose account of the De Soto expedition he summarized. We must therefore turn to this source in order to explain De Laet's map of Florida.

Herrera's position as official historian, which he held during the reign of three Spanish monarchs, gave him access to documents which he could not otherwise have secured. Thus the maps which he published in his *Descripcion de las Indias* are those of López de Velasco, and the text describing America is copied directly from a summary[5] of the latter's *Geografia y Descripcion Universal de las Indias*. To this summary Herrera added "some commentaries of an historical character, and considering the ideas of the time, he believed that by so doing he was rendering a service to the readers of his *Historia de las Indias*."[6] When he came to narrate the De Soto expedition, however, Herrera summarized the romantic and inaccurate narrative of Garcilaso de la Vega,[7] which is the least useful of all for what pertains to the geography of Florida; and De Laet either summarized Herrera's text or else translated it verbatim.

De Laet's conception of the Southern United States, as expressed by his map entitled *Florida et Regiones Vicinae*, was made widely known by Sanson's map of North America published in 1650.[8] A cursory comparison of these two maps makes it clear that Sanson transposed De Laet's Florida on a general map of North America, and also used De Laet's map as the basis of his special map of Florida of 1656.

On De Laet's map of Florida the Gulf Coast line is shown

& des parties d'icelle, comme de la Nouvelle France, Floride, des Antilles, Iucaya, Cuba Iamaica, . . . Amsterdam, 1619. No author is given, but the book is bound in with *Le Grand Rovtier de Mer, De Iean Hvgves de Linschot Hollandois . . . Nouvellement traduit de Flameng en François,* Amsterdam, 1619.

[5]Compare the text in *Descripcion de las Indias,* p. 20, with the summary published in *Colección de documentos inéditos,* XV, 442.

[6]*Historia General de los hechos de los Castellanos,* Madrid, 1934-1936, 5 volumes published, I, Proemio, lxxvi.

[7]*La Florida del Ynca. Historia del Adelantado Hernando de Soto, . . . escrita por el Ynca Garcilasso de la Vega,* Lisbon, 1605.

[8]*Amerique Septentrionale,* Par N. Sanson d'Abbeville Geog. du Roy, Paris, 1650.

only as far as the R. de Montañas; the name of this river, however, is not inscribed on the map. The remaining section of the coast line to Rio Pánuco is on another De Laet map in the *Nieuwe Wereldt,* entitled *Nova Hispania, Nova Galicia, Guatimala.* The nomenclature is that of earlier sea charts of the Gulf, but many legends are omitted.

The first of these two De Laet maps shows the hinterland up to latitude 38°, that is, up to about halfway between the mouths of the Ohio and of the Missouri. In the area bounded on the north by that line, on the east by the meridian of the east coast of the Floridan peninsula, on the south by the Gulf, and on the west by the Bahía del Espíritu Santo meridian, the course of thirteen rivers is drawn, seven of which empty directly into the Gulf, and six into the Bahía del Espíritu Santo; the latter bay being situated, as on all the early maps, in the northwest corner of the Gulf. Of these thirteen rivers, only three are named, R. de Flores, R. de Nieves, and R. del Spiritu Santo. This last river, however, does not empty into the bay of that name, but has its mouth slightly above latitude 29°, on the east side of Tacobago Bay, *i.e.,* Apalachee Bay, near the position of present-day Suwannee River.

As far as De Laet knew, there were two Bahías del Espíritu Santo in the Gulf of Mexico. Nearly all the maps which he could have seen indicated a bay of that name in the northeast corner of the Gulf, and the text of Herrera explicitly stated that De Soto landed at a bay called Bahía del Espíritu Santo, which was situated on the west coast of the Floridan peninsula. It would seem as though De Laet instructed his draughtsman, Hessel Gerritsz, to avoid duplicating the names of the bay and of the river. In his description of the coast, he wrote: "From Tampa to the Baye du St. Esprit, there are thirty leagues, and this bay is called Tocobaga; it is said to be at latitude 29°30'."[9] The earlier Dutch text has a third name for this bay: Bahía de Miruelo;[10] and in

---

[9] *L'histoire dv Nouveau Monde,* 130.

[10] "Ende noch soo veel [scl. thirty] leguen voorder [than the Bahia de Carlos also called Bahia de Iuan Ponce de Leon] de Bahia de Tampa, dry-en-dertych leguen von Tocobaga, die men anders noemt del Espiritu Santo, ofte de Meruelo, op de hooghte van neghen-en-twintich graden en een half"; *Nieuwe Wereldt,* 1630 edition, 167.

the following chapter, which is simply a copy of a passage in Huygen, there is a fourth name, Ancon Baxo,[11] for the eastern Bahía del Espíritu Santo.

At the time when De Laet was preparing the French edition of the *Nieuwe Wereldt,* this variety of names of the same bay and the identity of names for two bays at both corners of the Gulf must have appeared puzzling to him. He noted that from west Florida to the Rivière des Palmes, a distance of 300 leagues, there were few named landmarks:

> From Tocobaga, which seems to be called Ancon Baxo, to the riviere des Neges, commonly called Rio de Nieues, there are one hundred leagues; and from the riviere des Neges to the riviere des Fleurs, twenty leagues or more. From the riviere des Fleurs to the Baye du S. Esprit (which must be different from the preceding one) or Culata, as it is commonly known, there are seventy leagues. . . .[12]

This description of the coast is nothing else than that of Huygen—itself a direct translation of López de Gómara's text—in which De Laet inserted the alternative Tacobaga for Ancon Baxo, a new name which he read in Herrera who had it from López de Velasco. Besides, in the French text, De Laet failed to give the width of the Bahía del Espíritu Santo, though this was specifically mentioned as thirty leagues by both López de Gómara and Huygen.[13]

In view of all this, De Laet's map of Florida and the text which it illustrates can hardly be considered valid documentary evidence that the Mississippi was the Río del Espíritu Santo of the Spanish geographers. The Mississippi is certainly not the Río del Espíritu Santo which is marked as emptying into the Gulf on the east side of Apalachee Bay, and, of the six rivers which are shown emptying into the Bahía del Espíritu Santo in

---

[11]"De punkt van Florida heeft in de breedte twintich mijlen ende van daer tot Ancon Baxo; zijn hondert mijlen welck is gheleghen viijftich mijlen oost ende west met Rio Secco, welck de groote breedte is van Florida"; *ibid.,* 168.

[12]*L'histoire dv Nouveau Monde,* 130.

[13]Cf. López de Gómara, *Primera y segunda parte de la historia general de las Indias,* 1553², fol. viii; *Colección de documentos inéditos,* XV, 442; Herrera, *Descripcion de las Indias,* 20; Huygen, *Description de l'Amerique,* 7, 13; De Laet, *Nieuwe Wereldt,* 1630 edition, 167-168.

the northwest corner of the Gulf, not one is named. The Mississippi cannot be identified from the names of towns and provinces on the banks of one of these six rivers, for no towns are marked on five of them, and only along the easternmost river formed by two branches coming from the northeast did the mapmaker inscribe place names. On the banks of the eastern branch of this easternmost river, there appear the names of two towns, Aminoja [Aminoya] and Guacacoya [Guachoya], which we know were on the Mississippi, because Elvas has them on the Rio Grande and Garcilaso de la Vega has them both on the Chucagua. But we cannot conclude that the branch whereon they are here marked is the Mississippi, because on the western branch of this same river, De Laet has marked the four towns, Guaxule, Yachiaha, Acoste, and Coza, which we know for certain were hundreds of miles east of the Mississippi, not west of it as here.

De Laet quite realized how difficult it would be to draw a map of the interior by means of the various accounts of explorers to which he had access:

> The [French] and the [Spaniards] have given us such a sketchy knowledge of [Florida] (except insofar as the coast is concerned) and such confused data, especially with regard to the interior, that we are forced against our intention to make use here of the history of the events and of the routes followed by the explorers. For a knowledge of the Spanish expeditions we shall mainly make use of the Decades of Anthoine de Herrera.[14]

He then proceeds to narrate, in the following pages, the expeditions of Ponce de León, Ayllón, Narváez, and De Soto, according to Herrera's account.

The confusion spoken of is reflected on his map of Florida, the elements of which are derived from various quarters. Thus the delineation of the Floridan peninsula is taken from the map of Herrera in the *Descripcion de las Indias*; the delineation of the northern Gulf Coast is similar to that of the old Spanish sea charts, while its nomenclature is derived from López de Gómara and Huygen. The course of the rivers is purely imaginary, especially that of the six rivers emptying into the Bahía del Espíritu Santo. There is absolutely no basis either in Herrera or in any

---

[14] *L'histoire dv Nouveau Monde,* 103.

of the accounts of the De Soto expedition for this fanciful hydrography, and the Herrera place-names, as copied by De Laet in his text, are inserted at random. For instance, instead of indicating the westward march of De Soto's army by placing the Indian villages to the west of Tascalusa, the mapmaker inscribed them along a southwest-northeast line from Tascalusa to Capaha, and he inserted the names of the other western Indian villages within the ellipse formed by the plotted route from Hirrihigua to Tascalusa. Thus two of the westernmost towns, Guancane [Lacane] and Naguatex, are located due north of Apalachee Bay, instead of along the 94th meridian, which, at this latitude, means that the mapmaker placed these towns 600 miles east of their true position.

The reason for this original interpretation of De Soto's route is that Herrera's text is quite obscure with regard to directions, especially after the expedition reached Tascalusa. De Laet would have been able to give a better representation of the interior had he made use of the original text of Garcilaso de la Vega so poorly abridged by Herrera; but although he had the Inca's history of Peru, he does not seem to have had his *Florida*. Our present concern, however, is not to explain why the De Laet map is inaccurate, but simply to show that it affords no evidence by which the Rio del Espíritu Santo can be identified as the Mississippi.

All the details of the interior on De Laet's map of Florida were inserted in the map of North America published by John Blaeu in his atlas of 1642, and reissued in his atlas of 1653.[15] The basic map which he used is one of the earlier sea charts issued by his father, Willem Blaeu.[16] Since this basic map was different

---

[15]America Septentrionalis in *Nouveau Theatre dv Monde ou Novvel Atlas*, 3 volumes, Amsterdam, 1642, III, Part 2; and in *Nuevo Atlas*, Amsterdam, 1653, II. According to Phillips in Lowery, *A Descriptive List*, 139 and 146, this map was first published in the 1639 edition of Hondius' *Nouveau Théâtre du Monde*.

[16]The name of this distinguished family of map engravers and publishers was Janszoon; Jansz, is the abbreviated, Janssonius the Latinized form. In his "Descriptive Catalogue of Maps published separately by Blaeu," Wieder, *Monumenta Cartographica*, III, 69, wrote: "c. 1617. Paskaart van Guinea . . . Gedruct by Willem Jansz. Blauw inde Sonnewyser. . . . As the name 'Blauw' occurs on it, this must be the first occasion

from that used by De Laet, Blaeu's map contains fuller nomenclature along the Gulf Coast than is found on De Laet's, and for the same reason the smallest of the six rivers emptying into the Bahía del Espíritu Santo and nameless on De Laet's map is legended Rio del Espíritu Santo on Blaeu's map.

De Laet's conception of Florida exerted an indirect influence on the cartography of the Southern United States, because, as we have already mentioned and we shall now see in detail, this conception was embodied in Sanson's maps, specifically in those of 1650 and 1656, which had a very wide circulation.

On the earlier of these two maps the nomenclature of the coast of the Gulf of Mexico agrees with that found on the Blaeu map. East of the Bahía del Espíritu Santo, however, Sanson made two changes. He omitted De Laet's Rio de Nieues, and altered Blaeu's C. Escondido to "C. Escondido R.," thus making this cape both a cape and a river. This change is a trifle in comparison with the alterations observable on Sanson's 1656 map, which is truly a cartographical freak, as will be seen directly. On the 1650 map also, as on Mercator's planisphere, Florida is bounded by a mountain range running east-west for about fifteen degrees, then north-south down to latitude 30°, where the direction of the range is again toward the west. The inland nomenclature is that of De Laet, with an occasional omission, and with variations in the location of the Indian villages. With regard to

---

of his using it on a map; the first dated map which has 'Blau' added to the patronymic 'Jansz' is one of Holland dated 1621. It is still an open question when exactly Blaeu began to add the name Blaeu to his patronymic . . . The 1617 Globe has the patronymic only, whereas the augmented editions of those earlier maps all bear the name of Blaeu." The founder of this firm, Willem Janszoon, signed Blaeu only once, he generally spelled his adopted name Blauw or Blaeuw, but his sons and grandsons wrote Blaeu. The reason why Willem Janszoon added this name to his patronymic was to distinguish his firm from that of Jan Janszoon who opened a bookshop at Amsterdam after 1618. "Aangezien de winkels van Guilielmus Janssonius (Blaeu) en van Johannes Janssonius in elkanders nabijheid, op het Water stonden, en beider firma's zich voor een groot gedeelte op het gebied der aardrijkskunde bewogen, moest een en ander aanleiding geven tot vergissingen, die Blaeu, door de bijvoeging van zijn geslachtsnaam, schijnt te hebben willen voorkomen." P. J. H. Baudet, *Leven en Werken van Willem Jansz. Blaeu,* Utrecht, 1871, Bijlage I, Geslachtsnaam Blaeu, 117-119.

the Rio del Espíritu Santo, what we said in the analysis of the De Laet's map applies here also.

Sanson's map of 1656 includes the Southern United States, and is entitled *Le Nouveau Mexique et la Floride: Tirees de diverses Cartes et Relations.*[17] His indiscriminate insertion of geographical data "derived from various maps and relations" does little credit to his standing as a geographer and still less to his critical sense.

Not to mention duplication of names along the Gulf Coast and arbitrary positions of capes and rivers, the most astonishing features of this "up-to-date" map are the nomenclature of the interior and the hydrography of Florida, and these features result from the fact that he simply superimposed the Chaves map published by Ortelius in 1584 on the map of De Laet of 1625. It is quite evident that he did not study the relations mentioned in the title of his map and that the map is not "derived" from them at all. Sanson's aim seems to have been not to give as accurate a representation as possible of the area, but to publish a new map different from those then on the market. The following analysis shows how successfully he accomplished his purpose.

The mountain range encircling Florida is taken from Mercator's planisphere of 1569; the names of the two provinces north of these mountains, Calicuas and Capachi, are taken from the 1570 Ortelius map of the Americas, as is the legend "Suala M.," the western arm of the mountain range. This nomenclature first appeared on the Cock map mentioned in a previous section, whereon Monte Suala is the hilly country around Xuala or Xualla spoken of by the chroniclers of the De Soto expedition. For some unaccountable reason Cock placed these mountains fifteen degrees west of their actual position, and Ortelius' map repeats this error. Not content with copying this mistake, Sanson inserted another Xuala in the eastern part of Florida at about the place where this name appears on the De Laet map. Calicuas, as we saw when discussing Cock's map, is a variant for Coligoa or Coligua

---

[17]For all that pertains to Florida, there are but slight differences between this map and that entitled "La Floride," which Sanson published in his *L'Amerique en plvsievrs cartes,* Paris, 1657. The latter map is smaller and the anonymous engraver's skill was inferior to Somon, the engraver of the map of 1656.

of the early chroniclers, which is the same as Garcilaso de la Vega's Colima. Sanson was quite unaware of this fact; for, besides representing Colima as a town 150 miles due north of our Apalachee Bay, just as it appears on De Laet's map, he inserted Calicuaz as the name of a province, 700 miles northwest of the town.

More absurd still are the reduplications resulting from his superimposition of the Chaves nomenclature on the De Laet map. Thus while retaining Quigate at approximately the same position as on the De Laet map, he inserted Quigata seven degrees farther west as it appears on the Chaves map. Not realizing that Ychiaha of De Laet and Chiacha of Chaves were one and the same town, Sanson kept them both, at a distance of 150 miles from each other. He also retained De Laet's Acoste and Chaves' Coste; but his most ridiculous blunder is the reduplication of Naguatex. Sanson inscribed it on his map at about the same position as on the De Laet map, that is fifty miles west of Achalaque, four degrees north of the mouth of the Rio Grande; and then repeated it a second time as on the Chaves map. If we extend a perpendicular due north from the middle of the Mar pequeña, the nomenclature west of this line is exactly that of the Chaves map, and this nomenclature is duplicated east of this line because all the Chaves towns are situated there on De Laet's map faithfully copied by Sanson.

Further evidence that Sanson did not study the relations from which his 1656 map is allegedly derived appears from comparing his location of certain towns with their position as described in these relations. A few examples will suffice. According to all the chroniclers of the De Soto expedition, the towns of Tula, Coligoa, Lacane, and Autiamque (which Garcilaso de la Vega calls Colima, Guancane, and Utiangue, respectively) are located beyond the Mississippi. Sanson, however, placed them all ten degrees east of the Mississippi on a north-south line perpendicular to the mouth of the Rio Grande, which, according to him, empties into the Gulf at the western end of Apalachee Bay. Pacaha, an Indian village which the chroniclers located on the Mississippi between the Arkansas and the St. Francis rivers, was placed by Sanson about six degrees farther east, as on the

De Laet map, due north of our Cape San Blas, somewhere in Cleburne County, eastern Alabama.

Sanson's hydrography of the Southern United States is no less spectacular than his naming and placing of towns. After reproducing De Laet's network of rivers, he superimposed thereon all the western rivers found on the Chaves map. The result of this combination can easily be imagined. Next he undertook the business of rechristening these rivers. Along the northern Gulf Coast, reading from east to west, we find first, as in De Laet, a large Rio del Espíritu Santo emptying into the bay of Tacobaga, Apalachee Bay. Then comes a river called "S. M. de Ochus," a name given by some Spanish mapmakers to Pensacola Bay. After this there is an unnamed river, and finally a river which Sanson arbitrarily called Rio Grande, the name given to the Mississippi in the relations. This Rio Grande which empties into Apalachee Bay, as do the three other rivers just mentioned, is a combination of two nameless rivers which appear on De Laet's map, and in spite of its name, it is a small river in comparison with the other three. It is still more insignificant in comparison with the rivers to the west of it emptying into the Bahía del Espíritu Santo. There are six of these rivers, all left nameless by De Laet, and all but the westernmost of them have the same course as on De Laet's map. Sanson named two of them: the easternmost he calls Rio de Canaveral, which is Chaves' name for a river that empties directly into the Gulf,[18] for the westernmost he repeats the name Rio del Espíritu Santo. The southern course of this river is copied from De Laet; its northern course, which is not shown on De Laet's map, is taken from Chaves, and includes three large tributaries coming from the northeast and cutting across one of the two unnamed rivers that empty into the Bahía del Espíritu Santo. Even with these tributaries, this second Rio del Espíritu Santo of Sanson is much smaller than his Rio de Canaveral.

---

[18]Sanson was so anxious to make full use of his sources that, east of the Bahia del Espíritu Santo, near the mouth of another river left nameless by D. Laet, he repeated the legend R. de Canaveral. That this is the same river which appears at this place on De Laet's map may readily be seen from the delineation of its course as well as from the names of the two towns on its banks, Achussi and Tascalusa.

It does not seem necessary to prolong this list of cartographical "howlers." The interested reader can readily add many more of them by placing side by side the maps of Chaves, of De Laet, and of Sanson.

From what we have said, it should be clear that in the middle of the seventeenth century no European who consulted the maps of the most famous cartographer of the day could have been justified in concluding that the greatest of all rivers emptying into the Gulf on the northern coast was called Rio del Espíritu Santo. And anyone nowadays, either in Europe or in America, who carefully compares these maps with the sixteenth-century accounts of travels in the Southern United States, will almost certainly find it impossible to identify the Mississippi with the Rio del Espíritu Santo of the Spanish geographers.

We have already noted that De Laet's nomenclature of Florida is taken from the Herrera text, in turn a summary of Garcilaso de la Vega's narrative of the De Soto expedition. We also called attention to the obscurity of Herrera's text with regard to the route of the De Soto army, and the consequent difficulty of mapping this route through Florida. By means of the narratives of Biedma, Ranjel, and Elvas, taken in conjunction with their knowledge of the geography of the Southern United States, modern American students have been able to plot the route of the expedition with a fair degree of approximation. Geographers and mapmakers of the seventeenth century lacked not only our modern knowledge of this territory, but also the narratives of Biedma and Ranjel. De Laet's map, as we have seen, is ultimately based on an abridged version of Garcilaso de la Vega, and the Chaves map is the cartographical expression of an account which did not essentially differ from that of the Gentleman of Elvas. Sanson's 1656 map is therefore the only attempt made, prior to Claude Delisle's, to harmonize the geographical data of these two accounts. How successful was this attempt, the reader is now in a position to judge.

In the last decade of the seventeenth century Delisle used a French translation of the narrative of the Gentleman of Elvas, and a French adaptation of Garcilaso's narrative[19] for his drafts

---

[19]P. Richelet, *Histoire de la conquête de la Floride ou Relation de*

of the route of De Soto.[20] Of this adaptation it has been truly said: "To anyone at all acquainted with the courtly ponderosity of Garcilaso's production which is about two and a half times as long as the other three narratives combined, the condensation by Richelet, the only attempt to publish a translation in any language is little other than a work of genius."[21] A comparison between the Spanish and French texts, we may add, shows that in abridging the long-winded Garcilaso, Richelet exactly reproduced all the place names, directions, distances, and other pertinent geographical details.

Now we read in Garcilaso's original narrative that after leaving Alibamo, De Soto's army came upon a great river. This "because it was the greatest of all those our Spaniards saw in Florida was called by them Rio Grande and was not given any other name"; that is to say, it was called the great river par excellence, and was not called Rio del Espíritu Santo. He then goes on to give the Indian name of this Rio Grande: "Juan Coles [a member of the De Soto expedition and one of Garcilaso's three informants] says in his relation that this river is called Chucagua by the Indians."[22] And, let us note, nowhere in Garcilaso's narrative of the De Soto expedition is this Chucagua identified with the Rio del Espíritu Santo.

In 1674 a map was published in Paris which is the cartographical expression of Richelet's adaptation of Garcilaso de la Vega.[23] According to the title, the author of this map is the

---

ce qui s'est passé dans la découverte de ce Pays par Ferdinand de Soto, 2 volumes, Paris, 1670.

[20] Cf. "The Sources of the Delisle Map of America, 1703," in *Mid-America,* XXV, 1943, 280-283. The map showing the route of De Soto was published by Claude's son under the title *Carte de la Louisiane et Cours du Mississipi . . . Par Guill^(aume) Del'isle de l'Académie R^(1e) des Scien^(ces)*, Paris, 1718. The manuscript of this map is reproduced in Sara J. Tucker, *Indian Villages of the Illinois Country,* volume II, Scientific Papers, Illinois State Museum, Part I, *Atlas,* Springfield, Illinois, 1942, pl. XV.

[21] *Final Report* of the United States De Soto Expedition Commission, 8.

[22] *La Florida del Ynca,* f. 229. The Spanish text and Richelet's condensation of this passage may be seen in B. Boston, "The Route of De Soto: Delisle's Interpretation," *Mid-America,* XXI, 1939, 281, note 16; Herrera's summary of the passage will be found *ibid.,* note 17.

[23] *Amerique Septentrionale . . . Par le Sr Sanson, Presentee a Mon-*

"Sieur Sanson." It cannot be Nicolas Sanson, for he had been dead seven years when the map appeared, and there is good reason to believe that it was not drawn until after 1670, the year of the publication of Richelet's condensation, three years after Nicolas' death. For instance, in a map of North America published in 1669 under the name of Guillaume Sanson,[24] we find exactly the same nomenclature and hydrography of Florida as on the 1656 map which we analyzed above. The author of the 1674 map was therefore one of Nicolas Sanson's two sons, Guillaume or Adrien.

The question of its authorship is relatively unimportant. What matters is the fact already noted: that it is the cartographical expression of Garcilaso de la Vega's text as adapted by Richelet. With regard to the geography of Florida, although still inadequate and much inferior to the narratives of Elvas and Ranjel, this text is so much more complete than the Herrera summary that it enabled the mapmaker to avoid the glaring mistake of mislocating the Indian towns and villages as De Laet had done.

The basic map used by draughtsman is one of the numerous representations of the continent which were then on the market. Although we are more concerned with nomenclature, and especially with the hydrography of Florida than with its orography, a brief description of the orography on this map is given here, because of its influence on later explorers of the Mississippi Valley and on European theorists of its geography.

As on all the other maps made since the middle of the sixteenth century, Florida is surrounded by a mountain range, but with the following difference: instead of being completely shut off from Canada, there is a four degree gap southwest of Lake Erie, in the northern section of this range, between the "Apalache M." and longitude 288°. Beginning again at this point, the range fol-

---

seignevr le Davphin Par ... *Hubert Iailliot,* [Paris], 1674. Until 1670, Pierre Mariette had been the Sansons' publisher, but about this time, Hubert-Alexis Jaillot, a sculptor who had recently taken over his father-in-law's map publishing business, became acquainted with the Sanson brothers. Thereafter the latter entrusted to Jaillot the publication of their maps. Cf. F. Roland, *Alexis-Hubert Jaillot, Géographe du Roi Louis XIV (1632-1712),* Besançon, 1919, 11.

[24]*Amerique Septentrionale* ... *Revüe et Changée en plusieurs endroits* ... *Par G. Sanson,* ... Paris, 1669.

lows the fortieth parallel westward to longitude 273°, then turns south to the 27th parallel, whence it runs in a northwesterly direction to longitude 263°, and continues thereafter toward the north-northeast. The northern section of this range serves as the boundary between Florida and Canada; the western section separates Florida from Quivira, which latter province is indicated as a subdivision of New Mexico. From latitude 27°, at the point where the mountain range turns to the northwest, to latitude 25°, a dotted line along the 270th meridian marks the boundary between New Biscay and Florida, and another dotted line drawn eastward from this same point to the Gulf Coast separate Florida from the Province of Pánuco.

An examination of the rivers represented on this map will show beyond doubt that the Rio del Espíritu Santo cannot be identified with the Mississippi. There are two rivers on the map which correspond to the Rio del Espíritu Santo found on maps already analyzed. One, which is left nameless, flows due south from the Apalachee country into the Gulf on the east side of Apalachee Bay, just east of a river twice as long, which is marked "Rio Grande." This unnamed river is clearly the same as the Rio del Espíritu Santo of the 1656 Sanson map. The name "Rio Grande" is also borrowed from the Sanson map, though its course is not the same as Sanson's Rio Grande. The second Rio del Espíritu Santo on this map is one of four rivers emptying into the inevitable Bahía del Espíritu Santo, and is clearly legended. Of these four, the easternmost is named Chucagua, and its gigantic size gives evidence that it is indeed the Chucagua of Garcilaso de la Vega, the true Rio Grande of the Spaniards, and our Mississippi. In comparison with it, the river labeled Rio Grande is much smaller, and the Rio del Espíritu Santo is quite insignificant.

In his map of 1679, Duval makes it quite clear that the Mississippi was not known to him as the Rio del Espíritu Santo.[25] After combining the western part of the Sanson map of 1656 with the western part of the 1674 map, he legended "R. du St. Esprit" both the Rio del Espíritu Santo and the Rio de Canaveral of the 1656 map. Then, taking the upper course of the Chucagua

---

[25] *La Mer de Nort où sont la N^{le} France, la Floride, . . . Par P. Du-Val . . . ,* [Paris], 1669; reproduced in *Atlas brésilieu,* no. 79.

from the 1674 map, he marked it as "Chucagua R." and added to it the lower course of the Rio Grande of the 1656 map, inscribing the legend "R. grande de Chucagua" at its mouth, which occupies the same place on the Gulf Coast as does the mouth of the Rio Grande on the 1656 map.

VII
THE DISCOVERY OF THE MISSISSIPPI

If, as is commonly held, the Rio del Espíritu Santo is the Mississippi, and if, as is usually asserted, it was generally known in Europe and in America that the name of the great river flowing into the Gulf was the Rio del Espíritu Santo, it stands to reason that when the Mississippi was finally discovered by Jolliet from the North, and after it had been descended to the sea by La Salle, these two pathfinders, as well as the men versed in geography in New France and in Europe would have immediately concluded that the newly found river was the Rio del Espíritu Santo. Those who are most emphatic in asserting the existence of such prevailing knowledge base their assertion on a perfunctory perusal of a very few of the maps studied in the previous sections of this essay. However, from the almost dogmatic assurance of their pronouncements—which they modestly call demonstrations—one would think that they had at least analyzed the few maps upon which they dissert so glibly, and that they had ascertained their genesis or derivation.

In this section we shall consider: first, what was actually being said in New France about the great river when it was first heard of from the Indians; next, we shall see what was said of it after its discovery by Jolliet; in the third place, we shall examine whether La Salle, who descended the river to the sea, knew it under the name of Rio del Espíritu Santo; and finally, we must notice with what river people in France and in Canada identified the Mississippi after La Salle's voyage.

In general it can be said that those in New France who were interested in the geography of the North American continent derived their knowledge from the maps of Nicolas Sanson, for these maps immediately became so popular that imitations were everywhere in Europe, and the genuine maps themselves as well as Sanson's atlases soon found their way to Canada. It is known, for instance, that his 1650 map of North America was in Quebec before 1660, and there is positive evidence that his map of 1656 or a reissue thereof was to be found in Montreal in 1669.[1] When

---

[1] L. P. Kellogg, *Early Narratives of the Northwest 1634-1699*, New York, ᶜ1917, 204.

the Jesuit missionaries first began to hear of a large river west of the Great Lakes, they naturally tried to make this information fit in with what they knew of the country from these maps. Thus the geographical comments in the Relations of 1659-1660 and of 1661-1662 are a mixture of what was learned from the Indians about the south and the west, and of the data about these parts of North America represented on the Sanson map of 1650.

After mentioning that the Indians dwelling at the farthest end of Lake Superior had given an entirely new light on the geography of the West, the writer of the earlier of these two Relations goes on to say:

> Now, we know that proceeding southward for about three hundred leagues from the end of Lake Superior, . . . one comes to the *Baye du St. Esprit,* which lies at latitude 30° and at longtitude 280°, in the Gulf of Mexico, on the coast of Florida; and [we know that] from the same western end of Lake Superior, in a southwesterly direction, there are about two hundred leagues to another lake emptying, on the coast of New Granada, into the Vermilion Sea [Gulf of California], [which communicates with] the great Sea of the South [Pacific Ocean]. And it is from one of these two places that the Indians who live some sixty leagues to the west of our Lake Superior obtain European goods, and they even say that they have seen Europeans there.[2]

On the 1650 Sanson map of North America the 280th meridian crosses the thirtieth parallel in the "B. de Spiritu S^{to}." On the same map, "another lake" lying to the southwest of Lake Superior is shown emptying into the "Mar Vermejo" through the "Rio de Norte," and the distance between this lake and the western end of Lake Superior is about two thirds the distance between Lake Superior and the B. de Spiritu S^{to}. Again, the Mar Vermejo is shown washing the shores of "Nouvelle Grenade" and opening on the "Mer du Sud." Finally, the remainder of the above quotation—omitted because irrelevant here—in which the position of Port Nelson as well as the distance along the great circle between this point and Japan is given, removes all doubt that the author wrote the whole passage with the 1650 Sanson map of North America before him.

---

[2] R. G. Thwaites, ed., *The Jesuit Relations and Allied Documents,* 73 volumes, Cleveland, 1896-1901, hereinafter quoted as JR, 45:222.

We are not told whether "one comes to the Baye du St. Esprit" by way of a river; but this means of reaching the bay is specified in the second of the two Relations referred to above. On returning from one of their forays, Iroquois Indians had spoken of villages

> situated along a beautiful river which serves to carry people down to the great lake—for thus they call the sea—where they trade with Europeans who pray to God as we do, and who use rosary beads and bells for calling the people to prayer. From their description we judge them to be Spaniards. That sea must either be the *Baye du S. Esprit* in the Gulf of Mexico on the coast of Florida, or else the Vermilion Sea on the coast of New Granada, in the great South Sea.[3]

The author of this passage who also had the Sanson map of 1650, does not call the "beautiful river" Rio del Espíritu Santo for the simple reason that the hydrography of the Southern United States on this map is a mere copy of De Laet's Florida, and on the latter, as we saw, all the rivers emptying into the Bahía del Espíritu Santo are nameless.

Later in the sixties, as further information was received about the great river of the west, the Jesuits began to wonder where its mouth was situated. Father Allouez, who made its Algonquian name known to white men, conjectured that it discharged itself into the sea in the direction of Virginia.[4] Father Dablon, who in the autumn of 1669 had gone to the lower reaches of the Fox River, wrote as follows of the great river west of the Saint Francis Xavier mission in the Relation of 1669-1670: "It comes from the north and flows toward the south, and so far that the Indians who have descended it in search of enemies, after many days' journey have not found its mouth, which can only be toward the Sea of Florida [Gulf of Mexico] or that of California."[5] He makes no mention of a Baye du St. Esprit, but of the Gulf of Mexico in general, because Sanson's map, which he very probably had, showed Florida surrounded by a huge mountain range which would so deflect the course of the great river that it could

---

[3] JR, 47:146.
[4] JR, 51:46, 52.
[5] JR, 54:136.

not reach the Gulf at so high a latitude as that of the Baye du St. Esprit. He does suggest, however, that the river might flow around the mountain range, toward the southwest and so into the Gulf of California.

That mountain range impressed all who consulted these maps and they considered it a barrier which was bound to deflect the great river, either to the east, as Allouez thought, or to the west, as Dablon thought, and as Marquette himself believed in 1670. In a letter written after April 6 of the latter year, Marquette said: "It is hard to believe that the great river discharges its waters in Virginia; we think rather that it has its mouth in California."[6]

After his return to Quebec in 1671, Dablon wrote in the Relation of 1670-1671: "The great river which they call Mississippi . . . can only empty itself into the Sea of Florida."[7] Farther on in the same Relation, where he brings together all that had been learned from the Indians about the Mississippi, he says: "It seems to form an enclosure as it were of all our lakes, rising in the northern regions and flowing southwards to the sea, which we suppose to be either the Vermilion or the Florida Sea, because we do not know of any large rivers in that direction except those which empty into these two seas."[8] We may note that he makes no mention of a possible mouth in the Baye du St. Esprit, owing to the impassable mountain range shown on the map.

La Salle, as Galinée says in his narrative, joined the expedition of 1669 in the hope that by way of the Ohio he might find a "passage into the Vermilion Sea, into which M. de la Salle believed the Ohio River emptied."[9] That the Ohio was thought at the time to be the Mississippi is clear from Dollier's account of the journey of Courcelle to Lake Ontario in 1671:

> Two years ago, two ecclesiastics left here [to visit] diverse Indian nations, situated along a great river called by the Iroquois, Ohio, and by the Ottawa, Mississippi. Their design did not succeed . . . They learned, however, . . . that it was larger than the St. Lawrence River, . . . and that its

---

[6] JR, 54:188.
[7] JR, 55:96.
[8] JR, 55:206.
[9] Kellogg, *Early Narratives of the Northwest*, 168.

ordinary course was from east to west. After having closely examined the maps which we have of the coast of New Sweden, of the two Floridas, of Virginia, and of Old Mexico, I did not discover any river's mouth comparable to that of the St. Lawrence. This leads us to think that the river of which we speak disembogues into another sea—the determination of which I leave to the judgment of the more learned. Nevertheless, it is probable that it waters those countries in the direction of New Spain, which abound in gold and silver.[10]

Besides his narrative of the expedition of 1669-1670, Galinée wrote other memorials which are no longer extant. All that remains are five pages of geographical notes taken from them by Michel-Antoine Baudrand. The date of the memorials themselves cannot be precisely ascertained; we only know that they were written sometime in 1671. After describing the St. Lawrence, he continues as follows:

However, all the Indians of New France who have roamed over the country agree, on being questioned, that the St. Lawrence is not the most considerable river of Canada. They know of another one which the Iroquois call Ohio, and the Algonquian and Ottawa call Mississippi, as if to say "all the rivers in one," and it is called by them *the* great river, as though the St. Lawrence, which they all know, were in comparison but a small river.

This great river, he tells us, flows southwestward and Iroquois who have gone to its mouth have found the heat excessive,

It is difficult to say where this river empties: apparently not into the Sea of the North where no river like the Ohio disembogues, nor into Hudson Bay, for the Indians say that the weather is warm at its mouth, so that, it seems, it can only discharge itself into the Vermilion Sea.[11]

The position of the mouth of the Mississippi remained uncertain until the descent of the river by Jolliet and Marquette in June 1673. These explorers, it is true, turned northward while they were over 700 miles from the mouth of the Mississippi, but because Jolliet thought that he had reached latitude 33°, and be-

---

[10]E. B. O'Callaghan, ed., *Documents Relative to the Colonial History of the State of New York*, IX, Albany, 1855, 81.
[11]BN, Mss. fr., 15451:12-13.

cause of the direction of the river he was quite certain that the Mississippi emptied into the Gulf of Mexico.[12]

After an interview with Jolliet, Dablon wrote to France on August 1, 1674:

> The Father [Marquette] and the Sieur Jolliet have no doubt [that the Mississippi has its mouth] in the Gulf of Mexico, that is, in Florida. Its mouth cannot be in the east, in Virginia, for the lowest latitude of the Virginia coast is 34°, and they went as far down as latitude 33°, being then still fifty leagues from the sea. Nor [can the river have its mouth] in the west, in the Vermilion Sea, because their route which was always southward took them away from that sea. There remains, therefore, only Florida which is midway between both; and the Mississippi which our Frenchmen navigated, is very probably the river which geographers mark [on their maps] and call [rivière] du St. Esprit.[13]

As we have already seen, Dablon's statement that the Mississippi is "very probably" the Rio del Espíritu Santo cannot be justified by the maps at his disposal when he wrote this letter. On the other hand it is easy to see why he came to this conclusion. The reliability of the Sanson maps was so generally taken for granted that one would naturally attempt to reconcile all fresh information obtained from actual discoveries with the data contained in these maps. So long as Dablon had only the 1650 map, on which all the rivers emptying into the Bahía del Espíritu Santo are left nameless, there is not even a suggestion that the great river of the west is the Rio del Espíritu Santo. In 1674, however, he must have had either the 1656 map of Sanson or the 1657 variant thereof, on both of which two of the rivers which empty into this bay are named—the easternmost, R. de Canaveral, and the westernmost, Rio del Espíritu Santo. Since the first of these as shown on the map could not possibly be the river which Jolliet and Marquette descended, because of the position of its headwaters and the direction of its course, Dablon

---

[12]The terminus of the expedition is discussed in "Marquette's Autograph Map of the Mississippi River," *Mid-America*, XXVII, 1945, 41-44.

[13]Relation de la decouverte de la Mer du Sud . . . , Jesuit Archives of the Province of France, Fonds Brotier 155, Canada I, 2v. On this relation, cf. J. Delanglez, "The 1674 Account of the Discovery of the Mississippi," *Mid-America*, XXVI, 1944, 301-316.

would naturally assume that the Mississippi was "very probably" the second river named by Sanson.

The main point is, however, that this assertion of Dablon's cannot be taken as a proof that the Mississippi was known as the Rio del Espíritu Santo. For to say nothing of the cautious way in which he expresses his opinion, the evidence thus far presented is enough to show that he was mistaken.

In 1669, it will be remembered, La Salle thought that the great river which he had heard about from the Iroquois discharged itself into the Gulf of California. It is clear, moreover, from all the accounts written or dictated by him, as well as from the maps made under his direction, that he never completely rid himself of the idea that the great river was the Ohio.[14] This fact is apparent from Minet's map to be discussed later, which was an adaptation of La Salle's own map now lost,[15] as well as from the 1684 map made by Franquelin under the direct supervision of La Salle himself.[16] On both these maps, the upper course of the Mississippi, as also the Illinois, the Missouri, the Arkansas, and the Red River, are represented as converging tributaries or "rivières" which discharge into a large "fleuve" that runs off to the southwest into the sea.

Since Franquelin's map bases its delineation, nomenclature, and hydrography of the present-day United States on a variant of Coronelli's map, it represents the ideas of one of the best geographers in France at that time. The latter evidently did not believe that the Mississippi was identical with the Rio del Espíritu Santo, for both these rivers are shown on Franquelin's map, each with its own name and a course of its own: the Mississippi has its mouth on the eastern coast of Texas at latitude 26°, and the Rio del Espíritu Santo empties into the Baye du St. Esprit at the place where it is found on every map mentioned thus far in this essay.

---

[14]Cf. the "Chucagua" fragment examined below; the procès-verbaux of March 13-14, April 9, 1682, in P. Margry, *Découvertes et Établissements des Français dans l'Ouest et dans le Sud de l'Amérique Septentrionale*, 6 volumes, Paris, 1876-1888, II, 181 and 191; the narrative of Nicolas de la Salle, *ibid.*, I, 551; the map of Franquelin of 1684.

[15]*Some La Salle Journeys*, 34, note 51.

[16]"Franquelin, Mapmaker," *Mid-America*, XXV, 1943, 34 ff., 59-62.

The extant writings of La Salle show beyond the shadow of a doubt that he never identified the river which he had descended to the sea as the Rio del Espíritu Santo; and the reason is that he had never heard this name given to the great river discovered by the De Soto expedition. The two accounts of De Soto's expedition to which he could have had access at this time were the Elvas account, in which the great river of Florida is called the Rio Grande, and the account of Garcilaso de la Vega, in which the same great river is named the Chucagua. It is doubtful whether La Salle was acquainted with the original Portuguese Elvas narrative,[17] but he had a full knowledge of Garcilaso de la Vega's account, for he had with him, during his descent of the Mississippi in 1682, Richelet's adaptation of it,[18] as well as the Sanson-Jaillot map of 1674.[19]

A letter written by Tonti at Michilimackinac, July 23, 1682, tells of La Salle's first ideas about the location of the lower part of the river which he was descending: "We went below latitude 29°, leaving the Baye du S. Esprit on our left to the northeast. M. de la Salle believed that he was eighty leagues from the Santa Barbara mountains. He kept to himself the [exact] latitude of the mouth [of the Mississippi]. We traveled south and southwestward."[20] In his first memorial, written two years later, Tonti refers to the notion entertained by La Salle at an earlier stage of the voyage down the Mississippi, when they were at the level of the Taensa villages:

M. de la Salle had always believed that this river discharged

---

[17]See La Salle's letter to Father Oliva, April 5, 1666, published by G. J. Garraghan, in "Some Newly Discovered Marquette and La Salle Letters," *Archivum Historicum Societatis Jesu*, IV, 1935, 289. It was not necessary to know Portuguese to teach mathematics in Portugal during the seventeenth century, for this branch of the curriculum was taught in Latin. La Salle had some knowledge of Spanish, *ibid.*, 290.

[18]This is clear from the "Chucagua" fragment, *infra*.

[19]The two procés-verbaux referred to above do not make sense unless they are read in conjunction with this map.

[20]Tonti to ———, July 23, 1682, BN, Clairambault, 1016:168v. This letter is printed, French and English in parallel columns, in M. Habig, *The Franciscan Père Marquette, A Critical Biography of Father Zénobe Membré, O.F.M.*, New York, °1934, 215-229, the quotation will be found on p. 229.

itself into the Baye du Saint-Esprit; but when he had taken our latitude with his astrolabe, and found it to be thirty-one degrees, this led him to believe that we were in the Abscondido [Escondido] River, which was later found to be correct.[21]

As a matter of fact the expedition was never on the Rio Escondido, although La Salle himself later insisted that they were. This point, of importance in proving that he did not identify the Mississippi with the Rio del Espíritu Santo, is clearly brought out in the geographical digression contained in a letter which he wrote in the Illinois country a year after his voyage to the sea.[22]

In this letter, only a fragment of which is extant, he gives in detail the reasons why he believes the Chucagua of Garcilaso de la Vega is not the Mississippi, and why he is uncertain whether it flows into the river which he descended. Among these reasons, the following are relevant to our discussion here.

He argues that the Mississippi has its mouths facing east-southeast, and not due south; but the whole southern coast of Florida faces due south, except the portion between "the river marked Escondido on the maps and Pánuco. This Escondido is assuredly the Mississippi." He goes on to say that this same

---

[21] Margry, I, 602.

[22] This is the "Chucagua" fragment mentioned above; the beginning and the end are missing. The original is in BN, Clairambault, 1016:188-189v., printed in Margry, II, 196-203. Margry's editing of this La Salle autograph fragment is on a par with the editing of the rest of his compilation. There are misreadings, omissions, arbitrary paragraphings, and the punctuation of one passage quoted below renders the text unintelligible. J. F. Steward translated the defective Margry text in "La Salle a Victim of His Error in Longitude," *Transactions* of the Illinois State Historical Society for the Year 1910, XV, 129-132. There is no evidence whatever to support De Villiers' gratuitous assertion that La Salle himself mutilated the document, *L'expédition de Cavelier de la Salle dans le Golfe du Mexique,* (*1684-1687*), Paris, 1931, 34; nor is there any evidence for the still more gratuitous assertion of the following page; namely that, in France, La Salle "s'empressa de déchirer l'embouchure du Mississipi et le nom de Chucagoa accolé au fleuve Saint-Louis." How could La Salle have torn this map since it is a tracing made by or for Jean-Baptiste Bourguignon d'Anville who was born ten years after La Salle's death? Cf. "Franquelin, Mapmaker," *Mid-America,* XXV, 1943, 61-62.

portion of the coast is the only place which lies at latitude 27°, the latitude which he had observed at the mouth of the Mississippi. The rest of the Florida Coast, he remarks, is at latitude 30° at nearly every point.

There is, of course, another section of the coast where the mouth of the Mississippi might face east-southeast:

> There remains the Florida peninsula. But this is out of the question, for the Colbert [*i.e.,* Mississippi] River which at this latitude flows steadily eastward, or at most southeastward, would not have room in the east-west width of this peninsula, since it runs to the southeast for at least 120 leagues, from latitude 30° to latitude 27°, at which point it empties into the sea. This would be impossible in the width of the Florida peninsula. Now this is precisely the direction of the Rio Escondido. For this reason I maintain that we were near Mexico, and consequently in a river other than the Chucagua from which the Spaniards took such a long time to reach Mexico.

Besides explicitly identifying the Mississippi with the Rio Escondido, the above passage helps to explain his fundamental misconception of the geography of the Gulf, a misconception which was to have such fatal consequences a few years later. Like all his contemporaries, La Salle put implicit faith in the current maps of the Gulf of Mexico. Now prior to the maps of Delisle, no map of the period showed a delta jutting into the Gulf anywhere along the northern coast.[23] Since nothing but such a delta would explain the east-southeast direction of the mouths of the Mississippi, he would naturally locate his river on the Texas coast, even if he had not computed the latitude erroneously in 1682, an error partly due to his defective astrolabe.[24] Like the other members of his last expedition, he was on the lookout for the landmarks represented on the maps of the Gulf which they had with them, having no less confidence in the accuracy of these maps than he had in Garcilaso de la Vega's narrative, for, he

---

[23] Commenting on Delisle's map of 1703 in which the delta is shown, Joutel wrote: "Et sy ledit fleuve ce gette dans la mer a un cap aussy auancé que lauteur le marque il est a croire quon ne lauroit pas du manquer"; ASH, 115-9:no. 12.

[24] See Minet to Seignelay, July 6, 1685, Margry, II, 603.

wrote: "Assuredly the relation of Fernand Soto is not a fairy tale."[25]

In the letter from which we have quoted above he goes on to compare Garcilaso de la Vega's description of the Chucagua with what he had seen of the Mississippi, and then repeats his initial statement that the two rivers could not be one and the same, that the Chucagua must be farther east flowing parallel to the Mississippi. Next comes a restatement of what he had told Tonti at the Taensa portage:

> After we reached latitude 31°, all the Indians who go to the sea to make salt agreed in saying that the sea was situated to the east. Every morning we saw sea mists rising in the east and moving in toward the west, even against the wind. They were the mists [coming] from the shore of the Baye du St. Esprit and from the sea coast lying northeast-southwest from [Rio] Escondido to [Rio or Costa] de Pescadores and the above-mentioned Bay.[26]

Since the Rio del Espíritu Santo emptied into the Bahía del Espíritu Santo, and since La Salle clearly says that the mouth of his Rio Escondido—which is actually our Mississippi—is some distance below the bay, it is evident that this river is not identified by him with the Rio del Espíritu Santo.

After his return to France, La Salle altered the course of the Mississippi on his map, but he never wavered with regard to its identity. In Paris, when he gave a map to M. Tronson, he told this Sulpician that he had "entered the Gulf of Mexico not by way of the Baye du St. Esprit, but at latitude 27°, on the same meridian as Pánuco which is at the end of the Gulf, far beyond

---

[25] The "Chucagua" fragment, Margry, II, 197.

[26] "C'estoit les vapeurs de la coste de la baye du Saint Esprit et de la mer qui est entre Escondido, de Pescadores et ladite baye qui va du Nord-Est au Sud Ouest." There is a period at the end of this sentence, and the next sentence begins: "En passant ne doutés nullement. . . ." Margry edited this passage as follows: ". . . et de la mer qui est entre Escondido, Rio de Pescadores et ladite baye qui va du Nord-Est au Sud-Ouest en passant. Ne doutés nullement. . . ." He obviously did not understand what La Salle had written; and Steward's translation, *loc. cit.*, 131, of the second person plural of the imperative by the first person singular of the present indicative, "[I] do not doubt at all . . . ," makes things worse still.

this bay."[27] As we have already noted, Franquelin's map of 1684 shows the mouth of the Mississippi at exactly this point. Finally, in every letter written by Beaujeu and La Salle between October 1684 and March 1685, the Mississippi is said to be beyond the Baye du St. Esprit,[28] into which, as we have seen, the Rio del Espíritu Santo emptied on all the maps of the time.

Two passages in contemporary documents which seemingly indicate the opposite of this conclusion actually strengthen it when read in conjunction with a contemporary map. In Minet's letter to Seignelay accompanying the relation of the 1684-1685 voyage to the Gulf, the engineer says:

> I am sending you a map of Louisiana compiled from the best maps and memorials I could find. [On it] you will see that the river [Mississippi] and the lakes of Canada are drawn as on his [La Salle's] map. It is not certain that the river [Mississippi] empties directly into the lakes which we saw [the lagoons along the Texas coast]; it may just as well empty into the Baye du Saint-Esprit, and is perhaps the Rio Grande.[29]

The same idea is expressed in a letter which Arnoul, the Rochefort intendant, wrote to Seignelay the following day. Arnoul remarks that in the account of the 1684-1685 voyage which Beaujeu had shown him before sending it to Paris, no mention is made of the place where Beaujeu thought La Salle had landed on this occasion. He had told Arnoul that in his opinion the mouth of La Salle's river was farther east than Matagorda Bay:

> It may be situated in the direction of the Baye du St. Esprit or even below it [*i.e.*, still farther east], toward the Cape of Florida. This is all the more plausible because that section of America is nearest to Canada, and the English who live in Virginia where the *Joly* touched, say that behind the [Appalachian] mountains which surround their country there is a very large river.[30]

---

[27]Tronson to Belmont, April 15, 1685, no. 283; printed without date in Margry, II, 355.
[28]Cf. Margry, II, 522, 526, 529, 531, 532, etc., etc.
[29]Minet to Seignelay, July 7, 1685, AM, B 4, 10:296; printed in Margry, II, 603, under date of July 6. The original reads: "elle peut tomber aussy bien dans la baye du Saint-Esprit, peut-estre ses [*i.e.*, c'est] Rio Grande," and not as in Margry: "peut estre le Rio Grande."
[30][Arnoul] to Seignelay, July 8, 1685, BN, Mss. fr. n. a., 21331:307.

The above quotation is clearly a variant of what Minet wrote. The greater vagueness comes from the fact that Arnoul was less familiar with the maps of the period and was simply giving the substance of what Beaujeu had told him.

If the letter of Minet is compared with the data on his own map,[31] it can readily be seen that when he says "the river may just as well empty into the Baye du Saint-Esprit," the river which he means is not the Rio del Espíritu Santo, but the "R. Choucagoua,"[32] Garcilaso de la Vega's Chucagua. What he meant by the Rio Grande is just as obvious; for on his map, as on the Sanson-Jaillot of 1674, the river so named emptied into present-day Apalachee Bay. This latter river is also the one of which Arnoul, quoting Beaujeu, said that its mouth "may even be below," *i.e.*, farther east than "the Baye du St. Esprit, toward the Cape of Florida," as the peninsula was then called, according to many maps.

In view of all these texts, it should be clear that neither La Salle nor anyone else identified the Mississippi with the Rio del Espíritu Santo; and since the European maps from which they derived their knowledge of the hydrography of Florida contain, as we have seen, not the slightest evidence that the Mississippi was regarded as the Rio del Espíritu Santo, no one in either America or Europe made this identification.

There is, however, a contemporary document in which the Mississippi is explicitly identified with the Rio del Espíritu Santo. On November 8, 1682, La Barre, the recently arrived governor of New France, wrote from Quebec to Clairambault: "The Sieur de la Salle pretends to have discovered the mouth of the Mississippi." He adds that he had as yet received no news from the explorer himself, and promises to send more definite news when

---

[31]Carte de la Lovisiane, SHB, C 4044-4, reproduced in S. J. Tucker, ed., *Indian Villages of the Illinois Country*, pl. VII.

[32]On February 18, 1689, Beaujeu wrote to Villermont about a letter which he had seen. The sender of this letter said that Jean Cavelier, La Salle's brother, was coming to Canada with many complaints against Beaujeu; he also said that "La Salle was at the mouth of a river sixty leagues east of the place where I [Beaujeu] left him. This river must be the Chucagoa which empties into the Baye du Saint-Esprit, where the pilot of St. Augustine in Florida, whom he found at Petit-Goave, told us he had gone." Margry, III, 599.

he will have been longer in the colony than a mouth.[33] Yet, four days later, he wrote to the minister, this time:

> The Sieur de la Salle has written, by the Sieur de Tonti, that he actually discovered the mouth of the Mississippi . . . I am not making much of this discovery until I learn more about it, since this river is certainly the *Rivière de Spiritu Santo* in the Gulf of Mexico, at latitude 21°.[34]

This single text of a man who, as we know, was bent on belittling La Salle's achievements has little weight beside the mass of positive evidence which indicates the contrary. La Barre did not know what he was talking about, as is obvious from the latitude of the mouth of the *Rivière de Spiritu Santo*. The position which he gives, 21°, evidently a clerical error for 31°, is the latitude of the mouth of the Rio del Espíritu Santo on all the maps of the time. In 1682, La Salle had not communicated to anybody the latitude of the mouth of the Mississippi, and later he consistently said that it was situated on the 27th parallel or thereabouts. Moreover, as we have seen, in his letter of July 23, 1682, Tonti explicitly stated: "We went below latitude 29°, leaving the Baye du S. Esprit on our left to the northeast." It is very likely that on hearing of La Salle's discovery, La Barre tried to identify the river on one of Sanson's maps, as Dablon had done eight years earlier. But while the Jesuit made the identification "very probable," La Barre was positive that La Salle had descended the Rio del Espíritu Santo.

It cannot be said that the governor who had just arrived from France was voicing an opinion held at that time in that country, as the following facts will show.

After becoming acquainted with Peñalosa, Claude Bernou began to write memorials advocating the foundation of a French colony at the mouth of the Rio Bravo, our Rio Grande. In one of these memorials, composed in 1681 or 1682, among other advantages of such a settlement he mentions the fact that the colonists will

> be able to trade with the Indians and with the French who will settle in western Canada along the great Mississippi

---

[33] La Barre to Clairambault, November 8, 1682, BN, Clairambault, 448:315-316.

[34] La Barre to Seignelay, November 12, 1682, Margry, II, 302.

River which empties into the Baye du Saint-Esprit 100 or 120 leagues from the Rio Bravo; the course of the Mississippi is perfectly known to the Sieur de la Salle who is completing its exploration this year.[35]

The reason why Bernou has the Mississippi disemboguing into the Baye du Saint-Esprit is that, when he wrote the above memorial, he had seen maps made in Quebec which have the mouth of the Mississippi on the northern shore of this bay; but none of these maps labels the Mississippi "Rio del Espíritu Santo," and Bernou himself here does not identify it with the latter river.

In 1682, Bernou collaborated with Peronel in the making of a map of North America.[36] On this the course of the Mississippi is delineated down to latitude 36°, due north of the "B. de Sp. sancto." No Rio del Espíritu Santo is indicated as disemboguing into this bay, but a river of that name is shown emptying into Apalachee Bay.

At the beginning of the following year, 1683, after hearing of the completion of the discovery of the Mississippi by La Salle, Bernou wrote the so-called *Relation officielle,* a compilation of items culled for the most part from Membré's and Tonti's letters dated respectively June 6 and July 23, 1682, and from La Salle's letter of October of the same year. He combined this information with what he knew of the geography of North America as portrayed on available maps of that continent.[37] Moreover, it is clear that when he wrote this Relation, Bernou had before him a draft of the map which he had made in collaboration with Peronel the preceding year. The Mississippi, he says,

> empties into the Gulf of Mexico, beyond the Baye du Saint-Esprit, between latitudes 27° and 28°, at the place where some maps put the Rio de la Madalena [*e.g.,* the map of 1674] and others the Rio Escondido [*e.g.,* the map of Duval of 1679].[38]

---

[35]Proposition Pour établir une colonie dans la floride a l'embouchure du fleuue appellé Rio-bravo, BN, Clairambault, 1016:201. On these various memorials, cf. "Peñalosa's Expedition and La Salle," in *Some La Salle Journeys,* 66 ff.

[36]Cf. *Hennepin's Description of Louisiana,* 111 ff. The map is reproduced in *Indian Villages of the Illinois Country,* pl. VIII.

[37]Cf. "La Salle's Expedition of 1682," *Mid-America,* XXII, 1940, 3-37.

[38]An earlier map of Florida by Duval has, at latitude 27° 30′, "R.

It is about thirty leagues from the Rio Bravo, about sixty from the Rio de Palmas, and between ninety and one hundred leagues from Rio Pánuco where the nearest settlement of the Spaniards is situated.[39]

This *Relation officielle* was written at the time when Coronelli was making the Marly globe. At latitude 27° on this globe, in the band between the 275th and the 280th meridians, the Italian cartographer inscribed the following legend: "The mouth of the Mississippi River, . . . this spot is called on the maps Rio Escondido."[40] Four years later, in the description of the Mississippi in his *Atlante Veneto,* Coronelli wrote: "Some geographers quite erroneously legend Rio Escondido the mouth of this Colbert [Mississippi] River on their maps. There can be no doubt, however, that their mistake should by all means be corrected."[41] Later on he says: "Those geographers who had no knowledge of this river [Mississippi] have omitted to draw its course on their maps or have, quite mistakenly, placed there the Rio Escondido. We were the first to bring this fact to light."[42] Although Coronelli knew of the Rio del Espíritu Santo, he never identified it with the Mississippi. On his map of North America of 1688 and on the gores of the Florence globe, one of the mouths of the Mississippi is located on the east coast of Texas at latitude 27°, while another branch of the river has its mouth on the same meridian, one and a half degrees lower. On this map and on this globe, the "Bahia de le Espu S° ó Mar pequeno" opens on the northern coast of the Gulf five degrees east of the longitude of the mouths of the Mississippi, and at this point on the bay he has marked "Rio de lo Spiritu Santo."

Escondido o la Madalena," *Le Monde; ov, la Geographie Vniverselle,* Paris, 1663, between pp. 54 and 55. The maps in the two-volume edition of 1670 are the same as in that of 1663, but the text has been expanded. According to a note by Phillips in Lowery, *A Descriptive List of Maps,* 150, no. 153, the first edition—which I have not seen—was published in 1658. On the Sanson map of 1656 the R. Escondido and the R. de la Magdalena are two different rivers, the mouths of which are fifty miles apart.

[39] "Relation officielle," *Mid-America,* XXII, 1940, 33.
[40] Recueil des Inscriptions des Remarques Historiques et Geographiques qui sont sur le Globe Terrestre de Marly, BN, Mss. fr., 13365:64.
[41] *Atlante Veneto,* . . . 2 volumes, Venice, 1690-1691, I, 29.
[42] *Ibid.,* 132.

For those who "prefer to think" that these texts mean just the opposite of what they obviously say, and for those who choose to believe that the cartographers meant the exact contrary of what they inscribed in their maps the above evidence will not be conclusive. For those readers who have no such preferences, however, it supplies ample proof that the Mississippi was not known in North America as the Rio del Espíritu Santo by its two foremost explorers; and that no one in Europe who was expert in the geography of the New World identified the two rivers.

VIII

THE RE-DISCOVERY OF THE GULF COAST BY THE SPANIARDS

If there is any country where the name of the greatest of all rivers which empty into the Gulf of Mexico should have been known, that country is Spain. Spaniards had explored the Gulf more than a century and a half before La Salle reached the mouth of the Mississippi in 1682, and after 1686, when they named the Mississippi, they continued their search for the Bahía del Espíritu Santo into which, according to their maps, the Rio del Espíritu Santo discharged itself. This latter fact alone disposes of the gratuitous contention that the Mississippi was well known in government circles under the name of Rio del Espíritu Santo. Such an affirmation is certainly not based on cartographical or documentary evidence.

Twenty-five years ago, W. E. Dunn treated in detail of the rediscovery of the Gulf Coast by the Spaniards in a book which is still the only authoritative work on the subject.[1] In following his narrative, we shall call attention to those documents which indicate what knowledge of the geography of the Gulf of Mexico was possessed by the Spaniards in the last years of the seventeenth century.

In the first months of 1684, a Spanish pilot, Martín Echagaray,[2] presented to the King of Spain a memorial in which he proposed means to guard against the French menace in Florida, and offered to explore the coast from Apalache to Tampico. His knowledge of the coast, however, was very vague and was derived partly from a study of antiquated sea charts, partly from what he understood the Indians of Spanish Florida to have told him. From the port of Apalache to Tampico, according to his memorial,

> there are 300 leagues [of coast] ... In this region there are rivers of great volume (muy caudalosos). In particular the Bahía del Espíritu Santo is located there, 220 leagues from Vera Cruz and 140 leagues from the port of Apalache

---

[1]W. E. Dunn, *Spanish and French Rivalry in the Gulf Region of the United States, 1678-1702. The Beginnings of Texas and Pensacola*, Austin, Texas, 1917.

[2]On Echagaray, cf. Dunn, *op. cit.*, 20, note 19.

[present-day St. Marks, Florida]. Sea charts indicate this bay as being one of the best harbors. I learned from Indians who live on [the shores of] this bay and who are today living in the Apalache mission, that two very great rivers flow into it, one coming from a vast province called *Mouila,* it is not known where this river has its headwaters, but the other river comes from New Mexico.[3]

Like all Spanish pilots, Echagaray's distances are given in Spanish sea leagues, of which there are seventeen and one-half to a degree of latitude, approximately 3.9 miles to a league. By converting into miles the distance from Vera Cruz to the Bahía del Espíritu Santo, we see that the latter would be located on a modern map near the mouth of the Sabine River; and by converting into miles the distance from Apalache to the Bahía del Espíritu Santo, leaving out the delta of which Echagaray had no knowledge, the Bahía del Espíritu Santo would be in the vicinity of Calcasieu Pass, Louisiana. His description of this bay, however, applies to Mobile Bay, for it was from this vicinity that the Indians had come to the Apalache mission, and the two rivers which empty into his Bahía del Espíritu Santo are in fact the Alabama River which comes from the *Mouila* country, and the Tombigbee River, whose lower course comes from the northnorthwest, *i.e.,* from the general direction of New Mexico. His identification of our Mobile Bay with the Bahía del Espíritu Santo, although erroneous, is pertinent to the question at issue because, first, Echagaray does not name either of the two *rios muy caudalosos,* although on the map or maps of the Gulf which he had a Rio del Espíritu Santo was shown emptying into this bay; and secondly, if his Bahía del Espíritu Santo is Mobile Bay, it is clear that any river which he describes as emptying into it cannot be the Mississippi.

The main point of Echagaray's proposal with reference to the French menace was to transport to Florida fifty Spanish families from the Canary Islands and twenty-four Indian families from Campeche. The memorial was sent to the Consulado of the Casa de Contratación, but was set aside as being simply a scheme "to

---

[3]Echagaray Expediente, Archivo General de Indias (AGI), Mexico, 61-6-20. The quotations are from transcripts in the E. E. Ayer Collection of the Newberry Library, Chicago.

sell goods, free of all duties in Florida, Havana, and Campeche." Echagaray's appeal to the Council of the Indies late in 1684 was more favorably received, and his plan was approved by the *fiscal* whose report is dated April 11, 1685. The Junta de Guerra adopted this official's recommendations two months later, and the royal *cédula* in which Echagaray's offer was accepted was promulgated August 2, 1685.[4]

In his letter covering the memorial of 1684, Echagaray had advocated fortifying the Bahía del Espíritu Santo "so as to forbid its entrance to the enemy." Owing to the suspicions of the Consulado and the delay in taking action, the "enemy" had been in the place six months when the pilot's plan finally received the formal royal approval; for, in the preceding February, La Salle had planted his colony at Matagorda Bay. This disturbing news came to the Spaniards at the end of October 1685, through Denis Thomas, a member of La Salle's last expedition, who had deserted at Petit Goave. This Frenchman later embarked on a corsair which had been captured by the Spaniards and taken to Vera Cruz. The destination of the La Salle expedition, said the prisoner, was a place called *Micipipi*. To the Spaniards this name meant nothing at all, and although they surmised that the place referred to was somewhere on the Gulf Coast, four and a half years elapsed before they finally located the ruins of La Salle's settlement on the Garcitas River.

On hearing Thomas' account, Admiral Palacios at once consulted his maps and studied

> the unfamiliar region north of the Gulf of Mexico. There was clearly only one river leading from New France to the Gulf, along whose course one could travel for five hundred leagues. This was the river shown on the maps of the time as the "Rio del Espíritu Santo" flowing into the famous bay of the same name. Admiral Palacios . . . was not long in concluding that the probable site of the French settlement was on this river and bay. When he estimated the distance from Espíritu Santo to the various ports of Mexico and Florida, finding that it was only one hundred and twenty leagues from Apalache, one hundred and sixty-five from Vera Cruz, the dangerous proximity of the invaders was immediately realized.[5]

---

[4] Cf. Dunn, *op. cit.*, 25-29.
[5] *Id., ibid.*, 38.

The above is Dunn's résumé of Palacios' letter. We shall quote from the letter itself the prisoners' actual words as recorded by the admiral, in order to ascertain to what events of La Salle's voyages they referred; we shall next distinguish this information from Palacios' conclusions, and finally we shall see whether he identified the river descended by La Salle as the Rio del Espíritu Santo.

The prisoners said "that from New France . . . a Frenchman [La Salle] left with four or five other [Frenchmen] and some Indians and proceeded toward the west through an arm of the sea." The *brazo de mar que corre asia el occidente* is clearly the St. Lawrence River and the event referred to is the beginning of La Salle's explorations. "These men were inhabitants of a place in New France called Canada." The prisoners further said that after several months of navigation on this arm of the sea, La Salle and his men "came upon a very great lake, which they crossed on a ship to the end of it." This is an allusion to the *Griffon, la embarcazion,* and to the Great Lakes, *una laguna muy grande.* At the end of this laguna "they found another mouth [of a river] which they entered and after a voyage of more than 500 leagues they came upon a bay and a port at the entrance of which there was a small island dividing the river into two channels, and they descended these channels and reached the sea in the Gulf of Mexico." One readily sees that the expedition to the Gulf, 1681-1682, is made to appear as a continuation of the voyage of the *Griffon* in August-September 1679. The distance of more than five hundred leagues, it should be observed, is that from the *laguna muy grande* to the Gulf, and Palacios, like Echagaray and all other Spanish sailors, counted seventeen and one-half leagues for each degree of latitude. Consequently, since on his map the Bahía del Espíritu Santo was at latitude 30°, the mouth of the river at the end of the *laguna* was situated 1,200 miles north of the actual starting point of La Salle's voyage. The last words of the above quotation give the prisoners' version of the arrival of La Salle's party at the head of the passes of the Mississippi.

The fact that none of these men had been with La Salle, and that they were repeating, and probably further distorting what they had heard on the way from France to Santo Domingo, does not in the least affect the question, because Palacios' knowledge was wholly derived from what they told him, and it is on this knowledge that he calculated the position of the Bahía del Espíritu Santo where he thought La Salle had landed.

"When they were about half a league away from the sea the altitude of the sun was taken, but the prisoners declared that they did not know at what latitude they were; all that they know is that this [place] was called *Bahia de Misipipi*." La Salle then returned to France, where the expedition to the Gulf was organized. At Santo Domingo, on his way back, he bought what was necessary "to start a settlement *a esta Bahia llamada Misipipi*; this was about nine months ago." Here ends the information supplied by the French prisoners, which Palacios then tried to interpret with the help of a sea chart of the Gulf of Mexico.

> Considering the distance of 500 leagues over which they traveled, [the bay called Misipipi] can be none other than that which, on the maps, is called Bahía del Espíritu Santo, which, in the Gulf [of Mexico], lies at about latitude 30°, and is 145 leagues from a port named Tampico; from the said port to Vera Cruz there are 60 leagues, and from the said Bahía del Espíritu Santo to Mexico there are 280 leagues, partly along the coast and the rest overland.[6]

Comparing this direct translation of Palacios' letter with Dunn's résumé of it, we see that the latter inserted the name "Rio del Espíritu Santo" which is not mentioned at all. Palacios' reasoning is that the *bay* which the French called Misipipi can be none other than that which on his map is "la Bahia que llaman del espiritu ss.^{to}," situated at latitude 30°. It does not matter that, according to the prisoners, La Salle and his men came down to the bay by way of a river, for the question is, whether Palacios identified this river. Since he did not, we must conclude that the Mississippi was not generally known under the name Rio del Espíritu Santo; otherwise Palacios, an admiral of the Windward

---

[6]"Segun los maps en la distancia de 500 leguas que andubieron [the Bahia llamada Misipipi] no puede ser otra sino es la Bahia que llaman de espiritu ss.^{to} q esta en esta ensenada casi en 30 grados de Altura y dista de Vn puerto llamado Tampico 145 leg^s y desde el dho Puerto hasta dela Vera cruz 60 leguas y de la Bahia del espiritu ss^{to} a Mex^{co} ay 280 leguas seguidas parte por costa y los demas p. Tierra adentro," Palacios to ———, November 17, 1685, AGI, Mexico, 61-6-20.

Squadron, would surely not have hesitated to make the identification of the river and of the bay.

On October 27, 1685, Palacios wrote to the viceroy apprising him of what he had learned. When the courier reached Mexico City, November 3, an extraordinary council was immediately summoned, and two days later the viceroy issued a decree leaving everything in the hands of Palacios. The speed with which the Spaniards acted is a sufficient indication of their anxiety. Palacios had suggested that the search for La Salle's colony be made from Havana, because he thought the officials of that place were more familiar with the northern Gulf Coast. On November 21, 1685, a ship left Vera Cruz for Havana. On board were two pilots, Juan Enríquez Barroto, "an experienced draughtsman as well as a practical pilot," and Antonio Romero, who "had made many voyages from Havana to Apalache, and was personally familiar with that portion of the route to be followed." These two men had orders from the viceroy to the authorities of Havana commanding them to equip a vessel for the expeditionary voyage.[7]

One month after the arrival of the pilots at Havana the frigate *Nuestra Señora de la Concepción y San Joseph* left the habor for the north. From Apalachee Bay they sailed west, stopping at Pensacola Bay and at Mobile Bay. On March 4, 1686, "at eleven o'clock in the morning," wrote Jordán de Reina[8] in the only extant log of the journey,

> I reached the extremity of this shore along which were a number of tiny islands formed of mud flats through which poured a large river; the latter I called Palizada river because of the many stranded trees at its mouth; the water flows very swiftly. This was at latitude 29°3'.[9]

Dunn who supplemented this account with other documents adds that a prominent landmark in the vicinity of the mouth of the Mississippi was christened Cabo de Lodo and that a storm having arisen, the ship was driven to latitude 22°, whence they sailed to Vera Cruz, arriving there on March 13.[10]

---

[7]Dunn, *op. cit.*, 39-40.

[8]This log has been published in translation by I. A. Leonard, "The Spanish Re-Exploration of the Gulf Coast in 1686," *The Mississippi Valley Historical Review*, XXII, 1936, 551-557.

[9]Leonard, *loc. cit.*, 556.

[10]Dunn, *op. cit.*, 62.

Toward the end of the report from which we have already quoted, Jordán de Reina wrote as follows:

> I hereby state, Sir, that I believe this stream, to which I gave the name Palizada river, is a hundred leagues from the port of Apalachee; furthermore, I am of the opinion that the bay of Espíritu Santo which we were looking for does not lie on the parallel indicated on the charts; I believe that it is farther west, though not very much.[11]

These two passages from Jordán de Reina's journal clearly show that the Mississippi was not known to the Spaniards as the Rio del Espíritu Santo, otherwise the pilot would not have given it a new name, or at least would have said that this was the famous river which appears on so many maps. As can be seen, he specifically stated that the Bahía del Espíritu Santo into which the river of the same name flowed was farther west. When Barroto, Romero, and Jordán de Reina were at the mouth of the Mississippi, they were at the mouth of the largest of all rivers emptying into the Gulf on the northern coast. Whether they realized it or not is irrelevant, for the point is that the Mississippi is actually the greatest of all the rivers of Florida, as the Southern United States was known in those days, and it had been described as such by the chroniclers of the De Soto expedition two centuries earlier.

In view of Jordán de Reina's statements, the following passage from Dunn is somewhat surprising:

> In his report to the viceroy, Palacios stated that he believed that the expedition had approached very close to the French colony, for if the voyage had not been interrupted the Mississippi River and Espíritu Santo Bay would doubtless have been reached some thirty leagues west of the Rio de la Palizada (!).[12]

I have been unable to secure this report of Palacios to the viceroy, wherein he is said to have mentioned the Mississippi River, but I very much doubt if he actually mentioned it. We have seen above that Dunn inserted in another letter of Palacios a "Rio del Espíritu Santo," whereas in the letter itself there is no mention of such a river, but of a *Bahia de Misipipi,* which

---
[11]Leonard, *loc. cit.,* 556.
[12]Dunn, *op. cit.,* 62-63.

Palacios thought was the Bahía del Espíritu Santo shown on the maps which he consulted. According to Dunn himself, when the Spaniards were at the mouth of the Mississippi they little realized "that they had discovered the river for which they were seeking." As a matter of fact, the Spaniards were not looking for a river, but for the *Bahía* del Espíritu Santo where they thought that La Salle had planted his colony. Dunn continues:

> Their failure to recognize [the Mississippi River] as such, however, is not surprising. That great stream was supposed to empty into the excellent harbor of Espíritu Santo Bay; but no bay was to be seen, and a river whose channel was obstructed by débris was not imagined to be the one which La Salle would have chosen for the site of his settlement.

Here and elsewhere throughout his book, Dunn is taking for granted that the Rio del Espíritu Santo shown on the Spanish maps was our Mississippi. This is a postulate unsupported by any evidence whatever.

Shortly after the return of the Barroto-Romero expedition Palacios went to Spain, where he summed up the result of the first maritime expedition as follows:

> Today the whole coast of the Gulf of Mexico has been discovered and explored, with the exception of the strip from the mouth of the Rio de la Empalizada ... to that of the Rio de Tampico. In this distance of about one hundred leagues lies the Bay of Espíritu Santo, and west of it the Rio Bravo and other rivers which may form sand banks.[13]

As we have seen, Jordán de Reina thought that the Bahía del Espíritu Santo was west of the mouth of the Mississippi, "though not very much." At Vera Cruz Palacios interpreted this as "some thirty leagues." In order to determine approximately where the admiral located the Bahía del Espíritu Santo, this specific distance is not to be computed on that given by Jordán de Reina from St. Marks to the mouth of the Mississippi— one hundred leagues according to him—but on the distance given by Palacios; namely, that which he says intervened between the mouth of the Mississippi and Tampico—also one hundred leagues.

---

[13]*Id., ibid.,* 65, note 12, quoting the statement of Palacios, September 6, 1686, remitted by Oreytia to the Council of the Indies with a letter of September 28, 1686, AGI, Mexico, 61-6-20.

Thus according to Palacios, the Bahía del Espíritu Santo was situated about one third of the distance between the Rio de la Empalizada and Tampico, that is, between the Sabine River and Galveston Bay; and this again is approximately the location of the bay when its position on early Spanish sea charts of the Gulf of Mexico is transposed on a modern map.

The news of the French "invasion," as La Salle's expedition was called, reached Spain at about the time when Barroto and Romero returned to Vera Cruz from their exploratory journey. The authorities in Spain thought that the intruders had landed at the Bahía del Espíritu Santo, and remembering what Echagaray had written two years earlier, they judged that he knew the location of the famous bay and could guide a squadron to dislodge the French. Inquiries as to the present whereabouts of the pilot were made; he was finally found at Cadiz, and was ordered to come to Seville. In his letter of 1684 Echagaray had said that if more definite information were needed with regard to the location of the Bahía del Espíritu Santo, he was prepared to give it "according to a sea chart which is in my possession."

This sea chart[14] was sent to the chairman of the Casa de Contratación together with a letter dated April 20, 1686, which letter is nothing else than an explanation of the legends on the map.[15] The latter is a graphic representation combining what Echagaray had written in 1684 and what Palacios wrote in 1685 after questioning the French prisoners at Vera Cruz. Echagaray's own source of information for La Salle's voyage came from a Spanish boatswain who had been taken prisoner by the French. During the voyage from the Caribbean to France, the captors of this boatswain gave him their version of La Salle's journey to the Gulf in 1682. The *laguna* crossed by the *frances descubridor* for instance, stands for the Great Lakes. From the boatswain Echagaray also learned that La Salle sailed on this *laguna* "until he arrived at the mouth of the two rivers which flow out of the said *laguna* to the *baya del espiritu santo,* as can be seen from the dots" on the map.

---

[14]It has no title and is in AGI, Mexico, 61-6-20 (1), photograph in the Karpinski Series of Reproductions; a tracing from a photograph of the original is reproduced in Dunn, *op, cit.,* 44.

[15]Echagaray to Oreytia, April 20, 1686, AGI, Mexico, 61-6-20.

This dotted line, however, is inscribed not within the banks of one of the two rivers but within the banks of both, so that from the sketch it is impossible to know which one Echagaray thought was descended by La Salle. He very probably did not know. The two outlets of the *laguna* are possibly the two main routes taken by La Salle in his several exploratory journeys, scl., up the St. Joseph River and down the Kankakee and up the Chicago River to the Des Plaines. In his letter of 1684, it will be recalled, Echagaray had spoken of two rivers emptying into the Bahía del Espíritu Santo. It was only natural that he should combine what he had been told by the Indians with what he had learned from the boatswain. In fact, he repeats in his letter of 1686 the information about the two rivers flowing into the Bahía del Espíritu Santo, adding what the Indians had told him about the ignorance of the headwaters of these rivers. "Hence I infer that these two rivers originate in the aforesaid *laguna,* and that this is the route which the said *frances descubridor* may have taken to come down to the *baya del espiritu santo,* and it is my opinion that he could not have followed another route to the Gulf of Mexico where the *baya del espiritu santo* is, as indicated [on the map] by the letter 'V.'" On this map, the bay within which the letter "V" is inscribed has the following legend: "Baya de lo Sp" S'° Misipipi por otro nombre."

Here, therefore, just as in Palacios' report to the viceroy, it is clearly indicated that the *bay* which was called Espíritu Santo was the same as the *bay* which the prisoners told the admiral was named Misipipi. Echagaray gives no name to either of the two rivers followed by the *frances descubridor,* and does not call either of them Rio del Espíritu Santo. And yet, if the Mississippi had been known under that name, and had been as well known as it is claimed, Echagaray would have unhesitatingly called the river which La Salle descended, Rio del Espíritu Santo.

Dunn prefaces his brief description of Echagaray's map by saying: "An examination of this rude sketch . . . will show concretely the meager knowledge possessed by the Spaniards concerning the interior of the North American continent."[16] This statement is too general, it should read "possessed by Echa-

---

[16]Dunn, *op. cit.,* 45.

garay." It is indeed hard to believe that in the last quarter of the seventeenth century the geographers of the Casa de Contratación at least were so ignorant of the geography of the interior. Besides their own coastal manuscript maps, there were by that time scores of atlases and single printed maps everywhere in Europe, and most of the maps of Florida were ultimately based on maps made by former members of the Casa. Moreover, five years before Echagaray made this sketch, Thévenot had published a narrative of the Jolliet-Marquette expedition illustrated by a map showing the Mississippi from latitude 45° to the Gulf; on this map also, there are two tributaries of the Mississippi separated, by a short portage, from rivers emptying into Green Bay and into Lake Michigan. In 1683, that is, three years before the date of Echagaray's sketch, Hennepin had published his *Description de la Louisiane,* a book which was soon known all over Europe. In this book there was also a map of the Great Lakes which for all its defects gives an incomparably better picture of the St. Lawrence basin than the crude Echagaray drawing.

In September 1686, the new viceroy of Mexico, the Conde de Monclova, arrived at Vera Cruz with instructions from Madrid directing him to locate the French settlement and dislodge the intruders. He summoned the two pilots of the 1685-1686 expedition and learned from them of their fruitless search for La Salle's colony, which was thought to be planted on the Bahía del Espíritu Santo. Barroto and Romero were again chosen as pilots of the expedition sent out by Monclova. Sailing northward along the coast from Vera Cruz, they disregarded the names inscribed on their maps, and rechristened prominent landmarks; it was at this time that Matagorda Bay, for instance, was named San Bernardo. Continuing their journey northward, they finally reached the Rio de la Palizada. The same two pilots had thus explored the whole Gulf Coast from Vera Cruz to the tip of the Florida peninsula. Not having found any bay which fitted the description of the Bahía del Espíritu Santo, they concluded that the only bay which corresponded somewhat with it must be Mobile Bay, where they arrived May 22, and spent three days exploring its waters.[17]

---
[17] *Id., ibid.,* 75-77.

The next two paragraphs contain our comments on Dunn's conclusions regarding this identification, "No great river corresponding to the Mississippi or Rio del Espíritu Santo was found flowing into [Mobile Bay]." As we have already seen, the identification of the two rivers is purely gratuitous, for it supposes what should be proved. In this expedition as in the first, the Spaniards were not looking for a river, great or small, but for a bay where they expected to find the French settlement. On their way up the coast they had examined Matagorda Bay and had found the wreckage of a vessel which, they were convinced, had belonged to the French colony. "Instead [in Mobile Bay] six small streams were found, which could not be navigated even by such small boats as the pirogues." This would seem to indicate that the northern part of the bay was not explored. "In spite of the absence of a large river, however, the Spaniards concluded that they must be at the bay which was shown on the maps as Espíritu Santo." There was no Bahía del Espíritu Santo on the maps where Mobile Bay is, but a bay of that name is shown 500 miles farther west. The pilots' identification is partly based on *a priori* premises; namely, that La Salle must have planted his colony on a bay fit for settlement. As a matter of fact he had landed at one of the least suitable spots of the Gulf Coast, where, as Dunn observes on the preceding page, "the region was so low and swampy that the Spaniards seem to have been convinced that no sane person would attempt to settle there."

"No other body of water [except Mobile Bay] offered any inducements for settlement; . . ." Yet Pensacola Bay, only forty miles away, was much better suited than Mobile Bay both with regard to settlement and with regard to harbor facilities. ". . . Or corresponded so closely to the general description that had been given of Espíritu Santo Bay." This description, whatever it may have been, was derived from maps; and if this description applied to a bay on the northern coast, it corresponded much more closely, both regarding shape and position, to Galveston Bay than to Mobile Bay. Furthermore the characteristic feature in any description of the Bahía del Espíritu Santo is that on the northern shore of this bay a river of the same name flows into it. Now, the fact that the explorers only mention six small rivers which could not be navigated by their pirogues, shows that they did not find this characteristic trait. Again, the famous bay is described as an excellent port; but in sounding the entrance of Mobile Bay they only found between twenty and twenty-two spans of water, which even for seventeenth-century ocean-going vessels, would not make an excellent harbor. "In this uncertain fashion was the long-sought-for bay identified, but, no doubt, with entire correctness." Actually, the correctness of this identification is more than doubtful. To say nothing of its difference in shape and its position on the northern coast line, as well as the absence of a large river emptying into it, and the shallow entrance of this bay, there is the fact that these pilots could not have "recognized" this bay, because, as we have seen, they did not identify Mobile Bay with the Bahía del Espíritu Santo on their previous voyage thither.[18]

A quite probable explanation for this identification would seem to be the following. All that the Spaniards knew with regard to the position of the Bahía del Espíritu Santo was that their old maps located it in the northwest corner of the Gulf. When Barroto and Romero explored the eastern part of the Gulf Coast in their first expedition, they had assumed that the bay was situated west of the Rio de la Palizada. When they explored the western part of the coast, not finding any bay corresponding to what they expected, namely, an immense bay, a small sea, *mar pequeña*, which was at the same time an excellent harbor, and not wishing to discard a cartographical feature a century and a half old, they called Mobile Bay Bahía del Espíritu Santo. Two years later, when Alonso de León finally found the ruins of La Salle's settlement on the Garcitas River, he was much more logical in calling Matagorda Bay Bahía del Espíritu Santo,[19]

---

[18]Scaife also identified the Bahia del Espíritu Santo with Mobile Bay. When he wrote his essay, however, he had no knowledge of the documents brought to light by Dunn. As we noted before, *supra,* xi, Scaife's main reason for denying that the Mississippi was the Rio del Espíritu Santo of the Spanish geographers is because, on their maps, the river is shown to empty into a large bay, and this, he rightly contends, cannot be said of the Mississippi in historical times. The texts of the early Spanish chroniclers which he uses to identify the Bahia del Espíritu Santo with Mobile Bay prove nothing at all; and the texts from the writings of the French explorers have no bearing on this identification.

[19]Cf. the itinerary of Alonso de León in H. E. Bolton, ed., *Spanish*

because ever since the time the Spaniards heard of the coming of the French to the Gulf, they were convinced that they had landed at a bay of that name. Yet Matagorda Bay certainly did not correspond to the general description of the Bahía del Espíritu Santo as it appeared on the maps. With regard to shape it corresponded much more closely than Mobile Bay; and its distance from the position indicated on the maps of the Bahía del Espíritu Santo is only one fourth as great as that of Mobile Bay.

Even if, against the evidence, one were to hold that Mobile Bay was the Bahía del Espíritu Santo, it is quite obvious, as we pointed out when discussing Echagaray's letter of 1684, that the Mississippi was not the Rio del Espíritu Santo of the Spanish geographers.

The search for La Salle's colony led to the occupation of Pensacola by the Spaniards. This episode is mentioned here, because during the preliminaries to the actual founding of Pensacola the Spanish explorers often refer to the Bahía del Espíritu Santo and to the Mississippi. The diaries of the pilots of the preparatory journeys to Pensacola Bay are lost, but we have the writings of Sigüenza y Góngora, who, at that time, had a better knowledge, both theoretical and practical, of the Gulf Coast than anyone in Europe and in America, with the probable exception of his pupil Barroto. Besides this fundamental evidence, there are other reports and letters which will be examined in the following pages.

The earliest of these writings is the so-called Pez memorial,[20] a paper based on the findings of Barroto, wherein its alleged author,[21] Andrés de Pez, advocates the occupancy of Pensacola Bay by the Spaniards. A good index of how well the geography of the Gulf Coast was "generally known" in Spain is to be found in the opinion of the Marqués de la Granja who, in the name of the Council of the Indies, rejected Pez' proposal to colonize and

---

Exploration of the Southwest 1542-1706, New York, 1930, 388, 399-400; the letter of Father Massanet, *ibid.,* 353, 362-363; the opinion of Pez and Barroto on De León's itinerary, W. J. O'Donnell, ed., "La Salle's Occupation of Texas," *Mid-America,* XVIII, 1936, 119.

[20] I. A. Leonard, ed. and transl., *Spanish Approach to Pensacola, 1689-1693,* Albuquerque, New Mexico, 1939, 77-92.

[21] Cf. Dunn, *op. cit.,* 147, note 2, and 177; Leonard, *op. cit.,* 24-25, and 95.

fortify Pensacola. The *fiscal,* who had reported favorably on the project, had added that "a consideration of the reasons which obliged the Junta to drive the French out of Espíritu Santo bay (when it was said that they had occupied it) moves me to feel that the same thing ought to be done at Pensacola Bay." To this the Marqués de la Granja answered:

> Both [projects], apparently, were formed without reflection on or knowledge of the real motives for that effort which grew out of the understanding that the bay called Espíritu Santo was less than a hundred leagues from Vera Cruz and an equal distance from Mexico city. This belief was also the reason for requiring an examination of that port, for, according to the position assigned it and placed on the sea charts, it lay outside of the gulf of New Spain and around Negrillo reefs. Consequently, consideration was given to the advantages it possessed over Vera Cruz as a port of entry and departure in any season without fear of the "northers" which, as is well known, cause great concern in both the vicinity of Vera Cruz and Mexico city. But the same arguments do not hold now because of the great distance between the two places; the bay is more than three hundred leagues away on the coast running north and south, and in a position as much out of the way as though it were in the innermost part of the inlet called Carlos bay.[22]

One need not be a professional geographer to be startled by the Marqués de la Granja's fantastic geography of the Gulf of Mexico. According to him the Bahía del Espíritu Santo was "around Negrillo reefs," that is, one hundred miles from the northern coast of the Yucatan peninsula. And this statement was made by a member of the Council of the Indies under whose eyes there was a map of the Gulf of Mexico made by the technicians of the Casa de Contratación. The bay spoken of at the end of the above quotation, is the Bahía del Espíritu Santo on the west coast of the Florida peninsula, where nobody in Mexico ever thought of going to find La Salle's colony.

The Marqués de los Vélez to whom the viceroy of Mexico, the Conde de Galve, had written in support of the plan for fortifying Pensacola, also gave his opinion on the matter. He repeated what the Marqués de la Granja had said with regard to the former

---

[22]Leonard, *op. cit.*, 103-104.

belief that the Bahía del Espíritu Santo was near Vera Cruz and Mexico, and that its position was outside the Gulf of New Spain and around Negrillo reefs. But he improved upon his colleague by saying that Pensacola Bay and not the Bahía del Espíritu Santo is out of the way "and in the innermost part of the inlet called Carlos bay." The whole question of driving La Salle out of the Gulf of Mexico, he wrote, arose from the fact that he was believed to be close to Vera Cruz and Mexico City.

> Now we know that the San Bernardo lagoon is more than 180 leagues by sea from the port of Vera Cruz, and that ships cannot anchor in it, judging by what the pilots who have been there again and again have observed, and by what Monsieur La Salle found by experience in wrecking his ships so completely that not even the smallest could get out. Consequently, this new matter has an entirely different aspect, especially in the proposal to fortify Pensacola bay, which is more than 200 leagues from San Bernardo lagoon and, therefore, 380 from the port of Vera Cruz, and more than 700 by land from Mexico city, judging by the reports of the viceroy and the logs of the pilots who sailed along the gulf of Mexico. So, any conquest attempted by land is impracticable even if the bay were occupied and fortified by the French (which is not the case), for there is a distance of 400 leagues from the outermost missionary posts in the provinces of New Mexico and Coahuila . . . to Pensacola.[23]

After the plan to occupy and fortify Pensacola Bay had finally received the king's approval, the viceroy of Mexico wrote to Charles II in answer to the objections raised in Spain against the project.

> I am obliged, Sir, to state that at no time could one believe that Espíritu Santo bay was 100 leagues from Vera Cruz and an equal distance from Mexico city, for on the maps, both ancient and modern, it was located on the coast of the gulf of Mexico running east and west, 240 leagues from Vera Cruz, and more than 450 by land from Mexico city; nor does the bay appear on the maps outside the gulf of New Spain and around Negrillo reefs, for the entire coast lying between Cape Apalachee [?] and Cape Cotoche [Catoche] falls within this gulf. Consequently, Espíritu Santo bay, which is the same as the one called Mobile bay by the reconnoitering expe-

---

[23] *Id., ibid.,* 115-116.

ditions recently completed, can not be outside of the Gulf. Mobile bay is well protected from "northers," but its entrance is too shallow for ocean-going vessels.

The Committee has confused this bay with San Bernardo lagoon, . . . and they have taken it for Espíritu Santo bay, which is quite different. The lagoon is at the western end of the gulf coast line running east and west, in the middle of which is Espíritu Santo bay. But the one that we and other nations have not known hitherto is Pensacola Bay, seventeen leagues east of Espíritu Santo, and the only inlet on the Gulf shore into which . . . ocean-going vessels can enter; it is also within the gulf of Mexico and protected from the "northers." And so since this bay is distinct from the Espíritu Santo bay found by the early discoverers, and also different from San Bernardo lagoon, which the Committee mistook for Espíritu Santo, our fears regarding the purposes of the most Christian king remain valid. . . . That [Pensacola Bay] is 240 leagues from Vera Cruz,[24] as the map on the secretariat of the Council plainly shows, and not 380 leagues as the Committee asserts, is a fact, for the distance from Vera Cruz to San Bernardo lagoon must not be added to that from the latter to Pensacola because the three ports form a triangle. No one would claim it necessary to go via San Bernardo lagoon in sailing to Pensacola or *vice versa;* therefore, the Committee did not make a proper calculation of the distance.[25]

A few comments and rectifications are in place here. This geographical information came from Sigüenza, and incidentally it shows that in Mexico people knew the geography of the Gulf

---

[24]It is said in the Pez memorial: "Two hundred and forty leagues northeast by north of Vera Cruz, at thirty degrees and a half north latitude and on the same meridian as Lagartos river on the coast of Yucatan, Captain Juan Enríquez Barroto found this bay of Pensacola," Leonard, *op. cit.,* 79. What the editor means, *ibid.,* 96, note, is not quite clear: "Pensacola bay was nearer twenty degrees north latitude from Vera Cruz." Vera Cruz is 19° N. latitude; Pensacola 30° 30′ N. latitude. "Considering the Spanish league about three miles long, the two hundred and forty leagues indicated by Sigüenza y Góngora were around a hundred miles shorter than the actual distance." Sigüenza speaks here of the Spanish sea league which was slightly more than 3.9 miles long; hence Pensacola was 940 statute miles from Vera Cruz in a northeast by north direction; the actual distance is 945 statute miles. Sigüenza's figure, 240 leagues, was obtained by laying the *tronco de leguas* on any seventeenth-century sea chart of the Gulf.

[25]Leonard, *op. cit.,* 125-126.

much better than did the members of the Council of the Indies or of the Junta de Guerra. The obvious reason for this more adequate knowledge is the fact that there were then in Mexico trained pilots who, during the previous years, had explored the Gulf Coast from the Florida Keys to Vera Cruz.

We have seen how the pilots of the second maritime expedition came to identify Mobile Bay with the Bahía del Espíritu Santo of the old maps. Until the year of his death, 1696, the viceroy repeated this identification in his letters, but the explorers and seamen who sailed along the northern coast, except when repeating in their reports what the viceroy had written to them, call it simply Mobile Bay.[26] The records do not bear out Galve's contention that the Junta de Guerra or the Council of the Indies confused Mobile Bay with San Bernardo. Later, after the authorities in Spain had received the report of Alonso de León and called Matagorda Bay Bahía del Espíritu Santo, they were repeating the identification of the governor of Coahuila.

The viceroy's assertion that Espíritu Santo is in the middle of the northern coast of the Gulf goes counter to the position of this bay as it appears on every map drawn since the late twenties of the sixteenth century. We can readily admit that Pensacola Bay is distinct from the Bahía del Espíritu Santo of the early discoverers; as a matter of fact, these two were never alleged to be the same; what is not admissible is that Pensacola Bay was not known to the early discoverers, for Sigüenza himself declares that it was known to them as *Achussi*.

In pursuance of the royal decree of June 26, 1692, authorizing the exploration of the northern coast so as to find a suitable place for a settlement, an expedition was organized. The most important document written on this occasion is the description of the northern coast of the Gulf from Pensacola to the mouth of the Mississippi by Don Carlos de Sigüenza y Góngora.[27] His

---

[26]On none of the Spanish manuscript maps of the Gulf of Mexico which I have seen, is Mobile Bay called Bahia del Espíritu Santo; this name is given to Matagorda Bay and alternates with that of Lago de San Bernardo, the name given to this bay in 1687.

[27]Cf. I. A. Leonard, *Don Carlos de Sigüenza y Góngora, a Mexican Savant of the Seventeenth Century,* Berkeley, California, 1929; a brief sketch by the same author will be found in *The Mercurio Volante of Don*

instructions ordered him to make plans of the various bays and a map of the coast.[28] Of this cartographical work only the plan of Pensacola has survived.[29] "As the map of the coast line from Pensacola Bay to Mobile and from there to the mouth of the Mississippi, if Sigüenza y Góngora actually drew one, has not come to light, it is impossible to identify precisely many of the landmarks mentioned."[30] While this is regrettable, it has no bearing on the object of our inquiry.

The doubt expressed in the above quotation is quite reasonable. It is not certain that Sigüenza drew a map of the coast line, because it appears from his journal that he used a map made on previous expeditions, probably one of his pupil Barroto, as can be seen from a comparison of the nomenclature in the journal of Jordán de Reina with that in Sigüenza's journal and in his letter of June 16, 1699.[31] "The nearest contemporary maps available are those of French origin, and the geographical features of the region bear the names which Frenchmen bestowed upon them."[32] These Frenchmen, however, had copies of the Spanish maps made between 1686 and 1693, on which are found quite a few place-names which appeared on the lost maps. By means of these French maps as well as of eighteenth-century Spanish maps, most of the fundamental nomenclature of the originals can be reconstructed; this reconstruction, however, interesting

---

Carlos de Sigüenza y Góngora, Los Angeles, 1932, 21-27; and in *Spanish Approach to Pensacola,* 38-43.

[28] Leonard, *Spanish Approach,* 153.

[29] The original is in AGI, Mexico, 61-6-21 (1), photograph in the Karpinski Series of Reproductions; a tracing from a photograph is printed in Dunn, *Spanish and French Rivalry,* facing p. 160; actual size reproduction in colors, Leonard, *Spanish Approach to Pensacola.* Two other maps by Sigüenza are in the same legajo, one showing the route of Alonso de León to Matagorda Bay in 1689, cf. "The Sources of the Delisle Map of America, 1703," *Mid-America,* XXV, 1943, 298; the other illustrating De León's itinerary from Monclova to the Neches River. Photographs of both are in the Karpinski Series of Reproductions; the second is reproduced in H. E. Bolton, *Spanish Exploration of the Southwest,* facing p. 370.

[30] Leonard, *Spanish Approach,* 191, note 55.

[31] Informe de Don Carlos de Siguenza, June 16, 1699, AGI, Mexico, 61-6-22.

[32] Leonard, *Spanish Approach,* 191, note 55.

though it may be, has no place here; but the following points are pertinent.

First, the Mississippi is called by the name given to it at the time of the first maritime expedition of 1686, and there is not the slightest suggestion that the Spaniards believed this to have been the Rio del Espíritu Santo shown on their old maps. Secondly, except when they repeat the Conde de Galve's instructions neither Pez nor Sigüenza ever call Mobile Bay Bahía del Espíritu Santo, a name to which the viceroy was so much attached. But even if they had, and even if they had drawn the conclusion that the river which empties into the northern part of this bay was the Rio del Espíritu Santo, this would not identify it with the Mississippi. Thirdly, where Sigüenza gives his reasons for beginning a settlement on the banks of the Almirante [Blackwater] River, he says:

> It may be inferred from the remarkable width and depth of this river that its source is at considerable height, and that it is an outlet of one of the Great Lakes formed by the St. Lawrence or Canada river. Whatever it is, in coming down from the hinterland one can have full use of the valuable resources in the various provinces through the easy navigation along this river. That was the only reason for Robert La Salle's explorations, journeys, and ultimate ruin. What he in his fancy sought to accomplish by the Colbert river, we ourselves possess with indeed more than legal right.[33]

When Sigüenza wrote this report, he had already seen the Mississippi. Since he speaks of the Blackwater River as though it were comparable to the Mississippi itself, we can readily understand how other Spaniards at an earlier date might have been greatly impressed by the size of the Trinity River. The fact that the Trinity River flows into a bay, and that this bay is very similar in shape to the Bahía del Espíritu Santo as represented on old maps, and finally, that the bay is located in the northwest corner of the Gulf, would be so many added reasons for identifying the Rio del Espíritu Santo with the Trinity River, and the Bahía del Espíritu Santo with Galveston Bay.[34]

---

[33]*Id., ibid.*, 182.

[34]In *L'expédition de Cavelier de la Salle dans le Golfe du Mexique*, 54, De Villiers writes as follows: "La véritable baie de Spiritu Sancto était incontestablement la baie actuelle de Galveston." He bases this positive

identification on the position and shape of the bay as it appears on a few early maps which he consulted, but above all on three passages of *La Cosmographie avec l'Espère et le Régime du Soleil et du Nord,* by Jean Fonteneau *dit* Alphonce de Saintonge, G. Musset, ed., Paris, 1904. Two of De Villiers' suppositions are erroneous; firstly, that Alphonce sailed in the Gulf of Mexico up to the northern coast; secondly, that the "anse du Figuyer" is the Gulf. His identifying the "grande rivière qui est toute plaine de baptures" with the Mississippi will not stand examination, for Alphonce's distance, forty leagues from Cape Sable along the west coast of Florida, would place the mouth of this "Mississippi" at the latitude of Charlotte Bay. To try to make sense out of Alphonce's text on pages 509, 513, and 514, is a hopeless task, as can be verified by transposing his data on a modern map. He speaks indifferently of a "baye du Sainct Esperit," and of a "goulfe du Sainct Esperit"; of an "anse du Figuyer," and of a "cap du Figuyer." We find on one of the Dieppe maps a "p. de figeras," on another a "p. des figeras," and on a third "las higueras"; while the Cabot map has "C. de igueras." All these legends are inscribed in present-day Gulf of Honduras. See the map entitled "L'anse du Figuyer," *Cosmographie,* 504. Alphonce evidently translated the Spanish word into the French.

IX

THE COMING OF THE FRENCH

In spite of all the good reports about the value of Pensacola, and in spite of the royal order of June 13, 1694,[1] instructing the viceroy to begin without further delay the occupation and fortification of the Bahía Santa María de Galve, as the bay had been christened by Sigüenza,[2] little effort was made to carry out the recommendations contained in the *cédula*. But shortly after the signing of the Treaty of Ryswick, September 20, 1697, the news reached Spain that Louis XIV was preparing to found a colony in the Gulf of Mexico and to occupy some port there, presumably Pensacola. This news shook the lethargy of the Spanish government as nothing else could have done.[3] Every means of forestalling the French was taken, and as is well known, the Spaniards won the race by two months.

When the royal *cédula* reached Mexico City, July 14, 1698, the viceroy, José Sarmiento de Valladares, Conde de Moctezuma, called in Sigüenza and Andrés de Arriola.[4] Both wrote memoranda on the situation, and that of Arriola was made official.[5] This report is of interest because it is the only one of all the official documents of the period in which the Mississippi is identified not with the Rio del Espíritu Santo but with the Rio Grande of the chroniclers of the De Soto expedition. Besides exploring the country around Pensacola, Arriola wrote, a close examination should also be made of the coast

> from Pensacola to Mobile Bay and from there to the mouth of the river which we call Rio de la Palizada and the French call Colbert, which is located thirty leagues or thereabouts west of Mobile Bay. For I am certain that this was the river which M. de la Salle came to find, and not finding it, went to the San Bernardino [*sic*] lagoon where he planted his colony.

---

[1] Dunn, *Spanish and French Rivalry*, 171.
[2] Leonard, *Spanish Approach*, 155-156.
[3] Dunn, *op. cit.*, 173-176.
[4] On Arriola, cf. I. A. Leonard, "Don Andrés de Arriola and the Occupation of Pensacola Bay," in *New Spain and the Anglo-American West*, 2 volumes, privately printed, Los Angeles, 1932, I, 81-102; also Dunn, *Spanish and French Rivalry*, 176, note 50.
[5] Dunn, *op. cit.*, 177-179.

[In 1695] I visited the mouth of this river which appeared to me of very great volume (muy caudaloso) and seemed to be that which was called Rio Grande by those who went inland with the *adelantado* Hernando de Soto, navigating it on ships made [on its banks] for more than 400 leagues before they reached the sea. An exploration should also be made of the Rio de diso [*sic*] mentioned in his report by Don Carlos de Sigüenza, which is called [Rio] del Almirante, ascending it as far as possible, since it is navigable; because it may be a branch of the [Rio] de la Palizada; and since it is rather large, I suspect that the French may now come to take possession of its mouth, for they have more information about it than M. de la Salle had. In my opinion, it is no less important to occupy the mouth of this river, if it should turn out to be navigable and could be fortified, because of the length of its course, which can be conjectured from the Relation written by Father Fray Louis Hennepin, who accompanied M. de la Salle in his explorations; for the river has its source in the lakes of New France.[6]

While Arriola was writing his report, Iberville was making ready for his expedition to the Gulf. An account of these preparations, of the voyage itself, of his contact with the Spaniards at Pensacola, and of his finding of the mouth of the Mississippi, March 2, 1699, would be out of place here.[7] What is pertinent is to determine whether the French in the last years of the seventeenth century knew if the great river had once been called the Rio del Espíritu Santo. We have seen with what river Bernou, Coronelli, and others who had a theoretical knowledge of the geography of North America, identified the Mississippi. These opinions are summarized in the letter which Claude Delisle wrote to Cassini[8] and which was published in the *Journal des Sçavans*, May 10, 1700. In this letter the geographer says that when he wished to determine the position of the mouth of the Mississippi he had no reliable data, no astronomical observations, and moreover, "the Mississippi River is not marked on any of the maps," with the exception, he continues, of Nolin's map. He then goes

---

[6] Ynforme de Dn Andrés de Arriola, July 16, 1698, AGI, Mexico, 61-2-22.
[7] Cf. G. Frégault, *Iberville le Conquérant*, Montreal, °1944, "L'exploration du Mississipi et la colonisation de la Louisiane," 264-301.
[8] On the authorship of this letter, cf. "The sources of the Delisle Map of America, 1703," *Mid-America*, XXV, 1943, 279.

on to say that in the 1680's "it was a much debated question among those interested [in geography] to know exactly where the Mississippi River emptied into the sea." The difficulty of determining where its mouth was situated increased after the return of Beaujeu from the Gulf of Mexico, because neither he nor La Salle had found it.

> Since on no map of the coast of Florida was there any river to which one dared apply what was being said of the Mississippi, there were some, and the late M. Thévenot was one of them, who maintained that it had no noticeable mouth and that it lost itself either in the interior or in the coastal lagoons; for it is certain that the coast of Florida is very low and that the soil deposits brought down by the river have formed along the coast several islands which will one day be joined to the mainland as has happened in many other places in the world.
> Others, especially M. l'abbé Bernou, maintained that this could not be, and that a river corresponding to the description hitherto given must have a large and deep mouth. Others still thought they saved appearances by saying that the Mississippi was the river which the Spaniards called Rio Escondido, and such was the opinion of Father Coronelli, as can be seen on the map which the Sieur Nolin, his engraver, has published.[9]
> On this map, the Mississippi River empties into the sea at the western end of the Gulf of Mexico.[10]

Here we have an account of the theories about the Mississippi which were current in France during the last twenty years of the seventeenth century, as set forth by the foremost geographer of the day. He does not give the least indication that he himself or any of his contemporaries thought of the Mississippi as the Rio del Espíritu Santo. On three Delisle drafts made in 1696,[11] the course of the Mississippi is based on the account of La Salle's

---

[9] This is probably the map published at Paris in 1689; see Lowery, *A Descriptive List of Maps*, 181-182, no. 192. For all that pertains to the course of the Mississippi, it is similar to the other Coronelli maps.

[10] "Lettre de M. Delisle a M. Cassini, sur l'embouchure de la riviere de Mississipi," *Journal des Sçavans*, May 10, 1700, 204. The differences in the wording of the printed version and of the manuscript draft of this passage, ASH, 115-10:no. 17 B, are negligible.

[11] Cf. "The Sources of the Delisle Map of America, 1703," *Mid-America*, XXV, 1943, 279.

expedition as related in Le Clercq's *Premier Etablissement de la Foy dans la Nouvelle France*. The coast line is taken from one of the many maps in print at the time. On two of these drafts the river is called "Mississipi," but on the third, he inscribed near its mouth the following legend: "Mississipi ou Grande riviere nomme par les Espagnols Rio Escondido."[12] As is to be expected, on the three maps there is a Rio del Espíritu Santo which empties into Apalachee Bay; and east of the mouth of the Mississippi we have the usual Baye du Saint-Esprit. A simplified version of these drafts of 1696 appeared on the gores of his 1700 globe. On his map of America published late in 1700, there is a "B. du St. Esprit" designating Breton Sound. Considering that on the late Delisle maps this name is never used, we must inquire why he inscribed it on the 1700 map in this part of the Gulf.

When Iberville was preparing his expedition he consulted old maps similar to those which Palacios had consulted ten years earlier. This is clear both from an anonymous memoir of 1698[13] and from the plan which he submitted to Pontchartrain under date of June 18, 1698:

> On leaving Santo Domingo, I shall sail [northward] to reconnoiter the coast, [landing] fifty or sixty leagues to the west of the Florida peninsula. I shall then follow the coast observing it well, especially the rivers, to the baye du Saint-Esprit where all my ships will gather. I shall enter this bay and shall carefully examine it to see whether the Mississippi empties into it.[14]

This Baye du Saint-Esprit, he thought, was only about one hundred leagues from the "Baye St. Louis," to wit, Matagorda Bay. If the Baye du Saint-Esprit was a good port, he would build a fort there; if, however, Matagorda Bay were a better port, he intended to fortify it instead of the Baye du Saint-Esprit.

There is no question of the Mississippi being the Rio del Espíritu Santo, even though Iberville thought the river might perhaps empty into the Bahía del Espíritu Santo. For, as we

---

[12] The Spaniards did not call the Mississippi Rio Escondido. As we have seen, La Salle is the first one who made this identification, which Coronelli made known to the world in his *Atlante Veneto*.

[13] BN, Mss. fr. n. a., 21393:209-209v.

[14] Margry, IV, 53.

have seen, both Beaujeu and Minet also considered this possibility, but the latter's map makes it quite clear that he refers not to the Rio del Espíritu Santo, but to the Chucagua; and Iberville himself, after returning from his first voyage, positively identified the Chucagua with the Mississippi.[15]

But as Iberville's preparations progressed, more recent maps of the Gulf embodying the results of its survey by the Spanish pilots of the various expeditions sent to find La Salle's colony reached the French. We called attention above to the manner in which manuscripts fell into the hands of nationals of other countries, and added that some maps were obtained in the same way.[16]

Among the booty found on a Spanish ship captured by the French in 1697 there was a new map of the Gulf. The captain of the French ship brought this map of Toulon, where Nicolas de la Salle, the namesake of the explorer, who had descended the Mississippi with La Salle in 1682, was living at that time. When he learned that preparations were being made to send ships to the Gulf he wrote to Pontchartrain, August 14, 1698, informing the minister that he had some knowledge of the mouth of the Mississippi, and that he had this map of the Gulf. In his answer, the minister asked for details of this knowledge and told La Salle to send with his report "the map which had been found on the ship of the vice-admiral of the Armadille, taken by the *Bon* in 1697."[17] Nicolas de la Salle hastened to comply with this request, and on September 3 sent his report together "with a copy of the map of the coast of the Gulf of Mexico. M. Patoulet, captain [of the *Bon*] has the original parchment."[18] We know that this map was a recent one, quite different with regard to nomenclature from those previously possessed by the French, because a "rivière de la Palissade"[19] was indicated on it, a name which was first given to the Mississippi by Jordán de Reina in 1686,

---

[15]"Le Chicagoua est le Mississipi," in the copy by Claude Delisle of the "Reponse de Mr. d'Iberville a une lettre de Mr. Toinard sur la cote de Floride et de la riviere de Mississipi," September 11, 1699, ASH, 115-10:no. 6.

[16]*Supra,* 31, note 43.

[17]Nicolas de la Salle to Pontchartrain, August 27, 1698, Margry, IV, 82.

[18]*Id.* to [*id.*], September 3, 1698, ASH, 111-1:no. 18.

[19]De Boissieux to ———, January 7, 1698, Margry, IV, 44.

the year of the first maritime expedition sent to find La Salle's colony.

On December 4, 1698, the *Badine,* commanded by Iberville, and the *Marin,* commanded by Surgères, arrived at Cap Français [Cap Haitien, Haiti]. Two weeks later Iberville wrote from this place: "I shall follow my first intention which is to search for it [the Mississippi] in the vicinity of the *baye de Lago de Lodo.*"[20] This name is a mistaken reading on the part of Iberville, for it does not appear on any Spanish maps of the period. Instead, the legend *Cabo de Lodo* appears on the Spanish maps after 1686 at the tip of one of the "fingers" of the delta,[21] and the expanse of water between the delta and the mainland to the north was called *Laguna de Pez*.[22] That the map which Iberville used actually had the legend Cabo de Lodo instead of Lago de Lodo, is evident from a letter which Chasteaumorant wrote to the minister after his return to France.[23]

On this map there was also a Rio de la Palizada, but not until the following February did Iberville begin to suspect that this rio was the Mississippi. In a letter which he wrote to Pontchartrain from Léogane on December 31, 1698, he says: "I heard nothing about the Mississippi here, nobody has any definite idea as to where it is, nor has anyone sailed along that coast; I shall, therefore, follow my original plan and sail to the coast of Florida."[24] In a second letter to Pontchartrain, which is in the form of a journal, under date of February 4, he says that he has decided to "follow the coast to the rivière des Palissades which is twenty-five or thirty leagues from Mobile [Bay]"; and a week

[20]Iberville to Pontchartrain, December 19, 1698, *ibid.,* 89.

[21]Dunn, *Spanish and French Rivalry,* 62.

[22]Cf. Dunn, *op. cit.,* 162, and the map in AGI, Mexico, 61-6-17, photograph in the Karpinski Series of Reproductions; tracing from a photograph in the original in Dunn, 163. Sigüenza wrote: "desde el Cabo de lodo a la mouila que distan entre si por Mar y camino derecho Como treinta y Cinco leguas se forma Vna grandissima ensenada en que se hallan esparsidos infinitos Caios que son isletas de barro Arena y Mucara de varios Tamaños entre los quales ai dos trechos grandes de mar limpio que son la ensenada de Barroto . . . [y] la laguna de Pez que es el otro pedazo de mar limpio," Informe de Don Carlos de Sigüenza, June 16, 1699, AGI, Mexico, 61-6-22.

[23]Chasteaumorant to Pontchartrain, June 23, 1699, Margry, IV, 104.

[24]Iberville to Pontchartrain, December 31, 1698, *ibid.,* 91.

later, after anchoring off Ship Island, he wrote that he was first going to explore the "bay," that is, the expanse of water formed by the Mississippi Sound and Lake Borgne where he then was, and afterward "sail along the coast to the rivière de la Palissade *which is the Mississippi,* between fifteen and twenty leagues from here."[25]

Iberville had arrived at the latter conclusion by piecing together the information which he had received from various sources. Besides the map received from Pontchartrain, he had obtained another one at Santo Domingo,[26] and in addition it is likely that he saw the map given to the brother of one of Chasteaumorant's officers there.[27] Moreover, a Spanish pilot, Juan Vicente, boarded the *Badine* at Petit Goave.[28] Iberville also knew of the information which Chasteaumorant had received from the notorious Flemish corsair Laurent De Graff,[29] from Martínez at Pensacola, and especially from the "piloto real" loaned by Arriola.[30]

The details received from this last source were of considerable importance. Martínez, who came on board the *François* with this pilot, was asked many questions, "especially about the Palizada, San Bernardo, the Rio Bravo, and Pánuco"; he gave his host "little information, beyond stating that all those places were very shallow and uninviting."[31] But before dismissing the "piloto real," Chasteaumorant asked him if he had any knowledge of the Mississippi.

---

[25]*Id.* to *id.,* February 11, 1699, *ibid.,* 99. The log of the *Badine* under date of February 27 has the following: "I left the ships . . . for the Mississippi, which the Indians of these parts call Malbanchya, and the Spaniards [rivière] de la Palissade," *ibid.,* 157.

[26]Cf. the log of the *Badine, ibid.,* 149, and ASH, 115-10:no. 17 Z.

[27]Chasteaumorant to Pontchartrain, June 23, 1699, Margry, IV, 110.

[28]Ducasse to *id.,* January 13, 1699, *ibid.,* 92; Dunn, *Spanish and French Rivalry,* 187.

[29]Iberville to Pontchartrain, December 19, 1698, Margry, IV, 88; Ducasse to *id., ibid.,* 92; logs of the *Badine* and of the *Marin, ibid.,* 135, 216, 217; Mouffle to [Pontchartrain], June 23, 1699, AC, C 13A, 1:147; Dunn, *Spanish and French Rivalry,* 187.

[30]Chasteaumorant to Pontchartrain, June 23, 1699, Margy, IV, 109.

[31]Dunn, *op. cit.,* 187. Cf. Chasteaumorant to Pontchartrain, June 23, 1699, Margry, IV, 108-109; Mouffle to *id.,* June 23, 1699, AC, C 13A, 1:150.

He told me that he had not, but that he had heard of a river which was called the river of Canada, and which was located beyond the San Diego Islands,[32] but there was practically no water at its mouth. The freshets had brought down a very great quantity of trees which had formed a kind of bar, so that there was not, he thought, more than one fathom of water at its entrance. Moreover, the current was exceedingly strong.[33]

The pilot was, of course, describing the mouth of the Mississippi, as the French found out a month later. On March 2, 1699, Iberville wrote in the log of the *Badine* that "its appearance made me realize that it was the rivière de la Palissade which appeared to me well christened, for its mouth . . . seems to be blocked by these rocks,"[34] namely, by the driftwood incrusted with mud. How he later came to the conclusion that this was the river which La Salle had descended, is outside the scope of this essay. The point here is that if the Mississippi had been "generally known" at that time as the Rio del Espíritu Santo of the Spanish geographers, the Spaniards from 1686 on, and the French after 1698, would certainly have referred to the Mississippi under this name. As can be seen from what has been said in this and in the previous sections, no one except La Barre identifies the Mississippi with the Rio del Espíritu Santo.

---

[32]From the journal of Sigüenza y Góngora it appears that these islands, "Cayos de San Diego," as well as others, had been named at the time of the previous maritime expeditions, and that they were entered on the map—probably made by Barroto—which Sigüenza had with him. The "Cayos de San Diego" were the string of islands and islets from present-day Petit Bois Island to Isle au Pied; cf. Mouffle to [Pontchartrain], June 23, 1699, AC, C 13A, 1:151. The map referred to in this letter is entitled "Plan de la coste de la Floride depuis le Cap St Blanc iusques aux Isles St Diegue," and is among the papers of Pierre Arnoul in BN, Mss. fr. n. a., 21399:377. Two variants of this map which were made by Ste. Marie, a garde-marine who took part in the 1698-1699 expedition to the Gulf, (Perinet to Pontchartrain, May 4, 1700, ASH, 111-1:no. 20), are in ASH, 138-6-2 and 3.

[33]Chasteaumorant to Pontchartrain, June 23, 1699, Margry, IV, 110.

[34]Log of the *Badine, ibid.*, 159. "The Spaniards were correct in calling it *la riuiere de la palissade;* the mouth is entirely fenced in with trunks of trees, petrified and hard as rock," R. L. Butler, ed. and transl., *Journal of Paul Du Ru,* Chicago, 1934, 4.

There is mention made by Iberville of a rivière du Saint-Esprit which appeared on a freak map sent to Pontchartrain in August 1699.[35] This map had been forwarded to La Rochelle by the minister so that Iberville might examine it. The latter, after studying it, concluded that this rivière du Saint-Esprit was probably the Apalachicola River, certainly not the Mississippi, for its mouth was between eighty and one hundred leagues east of the mouth of the Mississippi.[36]

The last appearance of a rivière du Saint-Esprit in the eastern part of the Gulf of Mexico is on the Delisle map of 1703, where present-day Apalachicola River is legended "Apalachicoli ou Hitanichi et R. du St. Esprit."[37] On the next Delisle map of the United States, which was published in 1718 by Guillaume Delisle, Claude's son, the same river is called: "Riviere des Chattaux nommee cy devant Riv. des Apalachicolis."

Similarly, the Baye du Saint-Esprit, the Bahía del Espíritu Santo, which appeared so constantly on the northern coast since the late twenties of the sixteenth century, does not appear on

---

[35] Pontchartrain to Iberville, August 19, 1699, Margry, IV, 334.

[36] Iberville to Pontchartrain, August 30, 1699, *ibid.*, 341-342. A copy of a similar map was apparently sent by Thoinard to Iberville. In his answer, the latter repeated what he had written to the minister, ASH, 115-10:no. 6. One of the reasons why any attention was paid to this map was because, joined to it, there was a letter saying that the English had established themselves at the mouth of this rivière du Saint-Esprit. When he arrived at Santo Domingo, in December 1699, after being told that the English had begun a colony in the "baye de Spiritu Santo" (Ducasse to Pontchartrain, October 29, 1699, Margry, IV, 357), Iberville concluded that it was "the baye de Carlos . . . between the Cape of Florida and the Apalaches." This baye de Carlos, he said, is situated thirty leagues south-southeast of the Apalache Spanish settlement, that is, south of present-day Suwannee River. "It is this river [emptying into the baye de Carlos] which the English call [rivière] du Saint Esprit," Iberville to Pontchartrain, December 19, 1699, *ibid.*, 359. The governor of Carolina asserted that Pensacola was the bay which was called Bahia del Espíritu Santo on the maps; and the Spaniards of Pensacola feared the designs of the English on "Espíritu Santo Bay, also called Ascension Bay, and Tampa by the natives," Dunn, *Spanish and French Rivalry*, 197-198.

[37] The reason why Claude Delisle inscribed these three names on his map of 1703 was because he was not sure which was the correct one, cf. his "Questions sur la Route de Soto," in ASH, 115-10:no. 17 X.

maps made after 1700.[38] The reason why it was still inscribed on this map is because of a passage in a letter written by Iberville after his return from his first voyage to the Gulf. In this letter, which is based on the log of the *Badine,* he says that, leaving the ships safely at anchor off Ship Island, he "resolved to explore in launches the environs of *Lago de Lodo.* This is the name which the Spaniards give to what on the maps is called the baye du Saint-Esprit."[39]

As we have pointed out, Iberville simply misread the legend. What he calls *Lago de Lodo* is nothing else than the Mississippi Sound prolonged into Lake Borgne,[40] and from where he was it must have appeared as a bay. We should also note that he identified this *Lago Lodo* with the Baye du Saint-Esprit in a letter —the only place in all his extant writings—written after his return to France, when he had old and new maps of the Gulf before him. On the old Spanish maps of the Gulf there was a Bahía del Espíritu Santo on the northern coast, but, as we have seen, it was situated 400 miles west of the Mississippi Sound; and on the new Spanish maps of the Gulf, the only bay that was so named was Matagorda Bay. The Spaniards could not have called *Lago de Lodo* Baye du Saint-Esprit, for the simple reason that the first legend does not appear on their maps.

Copies of the above-mentioned letter containing this erroneous identification were made in France by people who were interested in the geography of the Gulf.[41] Guillaume Delisle

---

[38]The Bahia del Espíritu Santo of the northern Gulf Coast appears on map made after 1700, when the draughtsman or the engraver used a model dating back to the days of the *Conquista;* or when they faithfully copied that of Delisle, as in the case of the map of America issued at Amsterdam in 1708. Most of those who copied the Delisle map of 1700, however, omitted all mention of a Bahia del Espíritu Santo on the northern coast.

[39]Iberville to Pontchartrain, June 29, 1699, Margry, IV, 118.

[40]The following passages make the identification certain, Iberville to Pontchartrain, June 29, 1699, Margry, IV, 123, and Claude Delisle's notes entitled "Route de la Riv. d'Iberville [Manchac] au fort des Bilocchi [Ocean Springs, Mississippi] et du fort à cette Riv.," in ASH, 115-10 :no. 17 Z.

[41]There is a copy among the papers of Father Léonard de Ste. Catherine de Sienne, BN, Mss. fr., 9097 :45-48; and another in the Jesuit Archives of the Province of France, Fonds Brotier 165, Canada II. The same letter, with slight variants, was sent the following day, June 30, to an anonymous

extracted therefrom the pertinent geographical data,[42] and his father, as we saw, inscribed "B. du St. Esprit" to represent Breton Sound on his map of 1700. But his father had misgivings with regard to this legend. In order to clarify obscure points, Claude Delisle drew up a list of questions "to be put to M. d'Iberville." Among these questions are the following: "What is *Lago de Lodo?* Is it Lake Pontchartrain?[43] Does it include St. Louis and Biloxi bays? Is its entrance very far from the sea? Is it what the other maps call baye du St. Esprit?"[44] Peculiarly enough, Iberville did not answer these questions nor those which pertained to his first voyage. Yet Claude Delisle himself could have found answers to some of them by merely consulting Iberville's logs which he or Guillaume had copied. However that may be, Delisle had made up his mind about this "B. du St. Esprit" on the northern coast before he met Iberville in person, August 2, 1703,[45] for on his map which had been on the market four months previous to that date, and indeed on all the drafts of this map from 1701 on, there is no trace of a Baye du Saint-Esprit on the northern coast of the Gulf. The only bay of that name is on the west coast of the Florida peninsula corresponding to Tampa Bay which had been christened Bahía del Espíritu Santo when De Soto landed there in 1539.

---

addressee; it is found today among the Bernou papers, BN, Clairambault, 1016:514-517v. The passage discussed in the text is the same in all these letters.

[42]"Extrait d'une lettre de la Rochelle du 2 [*i.e.*, 29] juin 1699, de M^r dIberville touchant la R. de Mississipi," ASH, 115-10:no. 6.

[43]On an anonymous map, ASH, 138bis-1-5, showing the Gulf Coast from Cape San Blas to Matagorda Bay and the course of the Mississippi to the Illinois River, there are several interpolations by a different hand from that of the author of the map. For instance, the legend "Lac de pontchartrain" of the original is crossed out and is replaced by "baye du S. Esprit." Whoever made this change clearly wished "to harmonize" the nomenclature with that on old maps of the Gulf; and one of his insertions giving the farthest point reached by Iberville in 1700 shows that he was only superficially acquainted with the second voyage to the Mississippi.

[44]"Questions à fe. à M. d'Iberville," ASH, 115-10:no. 17 Q.

[45]ASH, 115-10:no. 17 Y.

X
CONCLUSION

The aim of all historical studies is—or should be—to reach certitude. In the present case we are certain that the Mississippi was not the Rio del Espíritu Santo of the Spanish geographers. The next step should be to determine the identity of this famous river, which begins to appear on maps made in the twenties of the sixteenth century. With the documentation now available this second step is impossible.

We know that it must be a river which empties into one of three bays on the northern coast: Mobile Bay, Lake Sabine, or Galveston Bay. The first one, for reasons given in a preceding section, is the most unlikely of the three; Lake Sabine is closer than is Mobile Bay to the position of the Bahía del Espíritu Santo on the northern coast line, but its shape does not correspond to that of the Bahía del Espíritu Santo as shown on the early maps. There remains Galveston Bay. Its shape, as well as its position in the northwest corner of the Gulf, approximates more closely to the Bahía del Espíritu Santo than the others. If Galveston Bay is in fact the Bahía del Espíritu Santo, the Rio del Espíritu Santo would be the Trinity River.

To say, as De Villiers does, that the "true Bahía del Espíritu Santo was unquestionably Galveston Bay," we should know whether a Bahía del Espíritu Santo of the same shape and position was represented on the Seville padrón, and supposing that it had been so represented, we should further know whether it was exactly reproduced by the makers of the maps which have come down to us. Precisely because these two points are doubtful, we cannot have more than probability for the identification of Galveston Bay with the Bahía del Espíritu Santo of these maps.

Since certainty regarding the identity of the Rio del Espíritu Santo is impossible from the evidence now available, the conclusion here indicated as a probability is not without value, but this probability does not become certainty as a result of one's subjective conviction. Probability, or even doubt, though a less pleasing result of the study of the evidence is not a less worthy one. When ignorance is inevitable there is no disgrace in admitting it. There is such a thing as learned ignorance, the true *docta ignorantia,* which is the result of a study of the possible sources of knowledge, even when they do not yield the knowledge expected from them.

In the present case we have inquired into these sources in order to ascertain the modern equivalent of a river inscribed on hundreds of maps and named Rio del Espíritu Santo; and we have come to one certain conclusion; namely, the Rio del Espíritu Santo of the Spanish geographers is *not* the Mississippi. To some this negative result may seem disappointing; it is, however, the only conclusion justified by the evidence.

A more satisfactory result of the examination of the documentation, although only a by-product, is a better knowledge of the historical cartography of the Gulf Coast and of that part of Spanish Florida lying west of the eighty-fourth meridian.

BIBLIOGRAPHY

MANUSCRIPTS

In the following calendar are listed only those manuscripts quoted from, or referred to in this essay.

KEY

AC, Archives des Colonies, Paris.
AGI, Archivo General de Indias, Seville.
AM, Archives de la Marine, Paris.
ASH, Archives du Service Hydrographique, Paris.
BN, Bibliothèque Nationale, Paris.
BN, Mss. fr., Manuscrits français.
BN, Mss. fr. n. a., Manuscrits français, nouvelles acquisitions.

1682, July 23.—Tonti to ———, BN, Clairambault, 1016:165v-168v.

1682, November 8.—La Barre to Clairambault, BN, Clairambault, 448: 315-316.

[1683?].—Fragment of an autograph letter of La Salle, BN, Clairambault, 1016:188-189v.

[after 1683].—Recueil des Inscriptions des Remarques Historiques et Geographiques qui sont sur le Globe Terrestre de Marly, BN, Mss. fr., 13365.

[1684].—Proposition Pour établir une colonie dans la floride a l'embouchure du fleuve appellé Rio-bravo, BN, Clairambault, 1016:199-205v.

1684, [Feb.-March].—Memoir of Echagaray, AGI, Mexico, 61-6-20.

1685, July 7.—Minet to Seignelay, AM, B 4, 10:296.

1685, July 8.—[Arnoul] to Seignelay, BN, Mss. fr. n.a., 21331:207.

1685, November 17.—Palacios to ———, AGI, Mexico, 61-6-20.

1686, April 20.—Echagaray to Oreytia, AGI, Mexico, 61-6-20.

[1698].—Memorandum of considerations to be borne in mind in planning and organizing an expedition to the baye du St. Esprit, for the discovery of the mouth of the Mississippi, BN, Mss. fr. n. a., 21393:209-209v.

1698, July 16.—Ynforme de Dn Andrés de Arriola, AGI, Mexico, 61-2-22.

1698, September 3.—Nicolas de La Salle to [Pontchartrain], ASH, 111-1: no. 18.

1699, June 16.—Informe de Don Carlos de Siguenza, AGI, Mexico, 61-6-22.

1699, June 23.—Mouffle to [Pontchartrain], AC, C 13A, 1:147-153.

[1699, June 29].—Copy of the "Lettre de Mons. d'Iberville," BN, Mss. fr., 9097:45-48.

1699, June 29.—Extrait d'une lettre de la Rochelle du 2 [*i.e.*, 29] juin 1699, de M^r d'Iberville touchant la R. de Mississipi, ASH, 115-10:no. 6.

1699, June 30.—Iberville to ———, BN, Clairambault, 1016:514-517.

1699, September 11.—Reponse de Mr. d'Iberville a une lettre de Mr. Toinard sur la cote de Floride et de la riviere de Mississipi, ASH, 115-10: no. 6.

[before May 1700].—Autograph draft of Claude Delisle's letter Cassini, ASH, 115-10:no. 17 B.

1700, May 4.—Perinet to Pontchartrain, ASH, 111-1:no. 20.
[1701-1703].—Notes by Claude Delisle on the coast of the Gulf of Mexico, ASH, 115-10:no. 17 Z.
[1702].—Questions à f[air]e à M. d'Iberville, ASH, 115-10:no. 17 Q.
[1703].—Questions sur la Route de Soto, ASH, 115-10:no. 17 X.
[1703].—Interview of Iberville by Claude Delisle, ASH, 115-10:no. 17 Y.
1703.—Joutel to Delisle, ASH, 115-9:no. 12.

ATLASES, BOOKS, COLLECTIONS OF MAPS, PERIODICALS

*Allard, C., pub., *Atlas Minor sive Tabulae Geographicae praecipuorum Regnorum Regionum, Insularum, Provinciarum, etc.*, Amsterdam, n. d.

Alfonce de Saintonge, J. Fonteneau *dit.*, *Les Voyages auantureux dv capitaine Ian Alfonce*, Poitiers, 1559².

*—————., *La Cosmographie avec l'Espère et le Régime du Soleil et du Nord*, G. Musset, ed., Paris, 1904.

Anania, G. L. d', *L'Vniversale del mondo*, Venice, 1596, reprint of the 1582³ edition.

*Anthiaume, A., *Cartes marines, Constructions navales, Voyages de Découverte chez les Normands*, 1500-1650, 2 vols., Paris, 1916.

—————., *Évolution et Enseignement de la Science Nautique en France, et principalement chez les Normands*, 2 vols., Paris, 1920.

[Apianus, P.], *Cosmographiae Introductio cum quibusdam Geometriae ac Astronomiae principiis ad eam rem necessariis*, n. p., 1533.

Apianus, P., *Cosmographicus Liber Petri Apiani Mathematici studiose collectus*, n. p. 1524.

*Apianus, P., and Gemma, R., *Cosmographia, siue Descriptio universi Orbis . . .*, Antwerp, 1584.

Atlas brésilien. See *Frontières entre le Brésil et la Guyane Française.*

Averdunk, H., and Müller-Reinhard, J., *Gerhard Mercator und die Geographen unter seinen Nachkommen*, in *Petermanns Mitteilungen*, XXXIX, 1915, Ergänzungsheft n. 182, 1914.

Avity, P. d', *Description Generale de l'Ameriqve Troisiesme Partie dv Monde. . . . Composé premierement par Pierre Davity . . . Novvelle edition, Reueu, Corrigé & augmenté, . . .* Par Iean Baptiste de Rocoles, Paris, 1660.

—————., *The Estates, Empires, & Principallities of the World. . . .* Translated out of French by Edw: Grimstone, . . . , London, 1615.

—————., *Les Etats, Empires, et Principavtez dv Monde, . . .* Par. le Sʳ. D.T.V.Y. . . . , Paris, 1615.

*Bagrow, L., *A. Ortelii Catalogus Cartographorum*, in *Petermanns Mitteilungen*, XLIII, 1929, Ergänzungsheft n. 199, 1928, and XLV, 1930, Ergänzungsheft n. 210, 1930.

*Bandelier, A. and F., eds., *The Journey of Alvar Nuñez Cabeza de Vaca*, New York, 1905.

An asterisk before the name of an author means that the book is referred to, or quoted from in this essay.

*[Barcia Carballido y Zuñiga, A. G. de], *Ensayo cronologico para la historia general de la Florida,* Madrid, 1723.

Basanier, M., comp., *L'histoire notable de la Floride, située es Indes Occidentales,* Paris, 1586.

*Baudet, P. J. H., *Leven en Werken van Willem Jansz. Blaeu,* Utrecht, 1871.

Benavides, A. de., *Memorial qve Fray IVAN de Santander . . . presenta a . . . Felipe Qvarto . . . ,* Madrid, 1630.

——————., *Reqveste Remonstrative av Roy d'Espagne svr la conversion du Nouveau Mexico,* Brussels, 1631.

Benzoni, G., *La Historia de Mondo Nvovo,* [Venice, 1565].

Berget, A., "Topographie. Les méthodes et les instruments du géographe-voyageur," *Revue de Géographie,* n. s. II, 1908, 511-560.

*Bertius, P., *Beschreibung der ganzen Welt,* Amsterdam, 1650.

*——————., *P. Bertii Tabvlarvm Geographicarvm contractarum Libri quinque,* Amsterdam, 1606.

*——————., *P. Bertii Tabvlarvm Geographicarvm contractarum Libri septem,* Amsterdam, 1616.

Bertolini, G. L., "Dell' uso della parola 'isola' per 'penisola' nell' alto Medio Evo," *Rivista Geographica Italiana,* XXXIV, 1927, 21-27.

Bion, N., *L'usage des globes céleste et terrestre, et des spheres suivant les différens systèmes du monde,* Paris, 1751[3].

*Blaeu, J., ed., *Le Grand Atlas ov Cosmographie Blaviane,* 12 vols., Amsterdam, 1663, vol. XII.

*——————., *Nouveau Theatre dv Monde ou Nouvel Atlas,* 3 vols., Amsterdam, 1642.

Blaeu, W. J., *Le Flambeau de la Navigation,* 2 vols., Amsterdam, 1620.

——————., *Gvilielmi Blaev Institvtio Astronomica De usu Globorum & Sphaerarum Caelestium ac Terrestrium:* . . . Latinè reddita à M. Hortensio, . . . , Amsterdam, 1640.

*Blaeu, W. J., and J., *Le Theatre du Monde; Ov Novvel Atlas . . . ,* 3 3 vols., Amsterdam, 1640-1643, vol. II; and 6 vols., in 4, Amsterdam, 1646-1650, vol. II, part 2.

Boissiere, C. de, and Gemma, R., *Les Principes d'Astronomie & Cosmographie auec l'vsage du Globe.* Le tout cōposé en Latin par Gemma Frizon, & mis en langage François par M. Claude de Boissiere, Daulphinois. Plus, est adiousté *l'usage de l'anneau Astronomic,* par ledict Gemma Frizon: *Et l'exposition de la Mappemonde,* composée par ledict de Boissiere, Paris, 1557.

*Bolton, H. E., *Spanish Exploration of the Southwest 1542-1706,* New York, 1930.

Bom, G. D., *Bijdragen tot eene Geschiednis van het Geslacht "van Keulen" als Boekhandelaars, Uitgevers, Kaart- en Instrumentmakers in Nederland: eene biblio-cartographische Studie,* Amsterdam, 1885.

*Boston, B., "The 'De Soto Map'," *Mid-America,* XXIII, 1941, 236-250.

*——————, "The Route of De Soto: Delisle's Interpretation," *Mid-America,* XXI, 1939, 277-297.

*Botero, G., *Le Relationi Vniversali di Giovanni Botero Benese divise en qvattro Parti,* Venice 1596, 1599; Spanish translation Valladolid, 1603; Italian in six parts, Venice, 1612; in seven parts, Venice, 1618; English translation, London, 1630.

*Bourne, E. G., *Narratives of the Career of Hernando de Soto in the Conquest of Florida,* 2 vols., New York, 1922².

Braun, G., *Zur Methode der Geographie als Wissenschaft,* Greifwald, 1925.

Bréard, C. and P., eds., *Documents relatifs à la marine normande et à ses armements au XVIᵉ et XVIIᵉ siècle pour le Canada, les Antilles, le Brèsil et les Indes,* Rouen, 1889.

Brevoort, J. C., *Verrazano the Navigator,* New York, 1874.

*British Museum, *Catalogue of the Manuscript Maps, Charts, and Plans,* 3 vols., London, 1844-1859.

*————, *Catalogue of the Printed Maps, Plans, and Charts,* London, 1885.

[Broë, S. de], ed. and transl., *Histoire de la conqueste de la Floride, par les Espagnols, sous Ferdinand de Soto.* Ecrite en Portugais par un Gentil-homme de la ville d'Elvas. Par M. D. C., Paris, 1685.

Brown, L. A., *Jean Domenique Cassini and his World Map of 1696,* Ann Arbor, Mich., 1941.

*Butler, R. L., ed. and transl., *Journal of Paul Du Ru,* Chicago, 1934.

*Byington, C., *An English and Choctaw Definer,* New York, 1852.

*Cabeza de Vaca, A. N., *Le relacion y comentarios del gouernador Aluar Nuñez Cabeça de vaca, de lo acaesido en las dos jornadas que hizo a las Indias,* Valladolid, 1555.

Camus, A. G., *Mémoire sur la collection des grands et petits voyages [de de Bry] et sur la collection des voyages de Melchisedech Thevenot,* Paris, 1802.

*Caraci, G., *Tabulae Geographicae Vetustiores in Italia Adservatae,* 3 vols., Florence, 1926-1932.

Carpenter, N., *Geography Delineated Forth in Two Bookes. Containing the Sphaericall and Topicall Parts Thereof,* Oxford, 1625.

Cartas de Indias, Madrid, 1877.

Chambers, H. E., *West Florida and Its Relation to the Historical Cartography of the United States,* Baltimore, 1898.

[Chatelain, H. A.], *Atlas historique; ou, Nouvelle introduction à l'histoire . . . & à la géographie. . . . Representée dans de Nouvelles Cartes,* 3 vols., Amsterdam, 1708.

Clüver, P., *Philippi Cluverii Introdvctionis in Vniversam Geographiam, tam Veterem quàm Novam, Libri VI,* Leyden, 1641; the text was expanded and maps were added in later editions, v.g., 1672, 1683, 1697, 1717, 1729.

Colección de documentos inéditos relativos al descubrimiento, conquista y colonización de la posesiones Españoles en América y Oceanía, 42 vols., Madrid, 1864-1884.

Colvin, S., *Early Engraving & Engravers in England (1545-1695). A Critical and Historical Essay,* London, 1905.

*Coote, C. H., *Autotype Facsimiles of Three Mappemondes,* privately printed, [Aberdeen], 1898.

*Coronelli, V. M., *Atlante Veneto, Nel quale si contiene la Descrittione Geographica degl' Imperij, Regni, Provincie, e Stati dell' Universo,* 2 vols., Venice, 1690-1691.

*——————., *Isolario dell' Atlante Veneto,* 2 vols., Venice, 1696-1697.

*Cortés, F., *Praeclara Ferdinādi Cortesii de Noua maris Oceani Hyspanica Narratio,* Nuremberg, 1524.

*——————., *La preclara Narratione de Ferdinando Cortese della Nuoua Hispagna del Mare Oceano,* Venice, 1524.

*Cortesão, A., *Cartografia e cartógrafos portugueses dos séculos XV e XVI,* 2 vols., Lisbon, 1935.

Corthell, E. L., "The Delta of the Mississippi River," *National Geographic Magazine,* VII, 1897, 351-354.

Crouse, N. M., *Contributions of the Canadian Jesuits to the Geographical Knowledge of New France 1632-1675,* [Ithaca, N. Y.], 1924.

*Dahlgreen, E. W., ed., *Map of the World by Alonzo de Santa Cruz 1542,* Stockholm, 1892.

Daly, C. P., *The Early History of Cartography, or what we know of Maps and Mapmaking before the time of Mercator,* New York, 1879.

*Dávila Padilla, A., *Historia de la Fvndacion y discurso de la Prouincia de Santiago de Mexico, de la Orden de predicadores, . . . ,* Madrid, 1596.

Davis, W. M., *Geographical Essays,* New York, ᵉ1909.

*Dawson, S. E., *The Saint Lawrence Basin, and Its Borderlands,* London, 1905.

*De Jode, G. and C., *Speculum Orbis Terrarum,* [Antwerp, 1593-1613].

*De Laet, J., *L'histoire dv Nouveau Monde ou Description des Indes Occidentales Contenant dix-huict Liures,* Leyden, 1640.

*——————., *Nieuvve Wereldt Ofte Beschrijvinghe van VVest-Indien,* Leyden, 1630².

*——————., *Novus Orbis, seu Descriptionis Indiae Occidentalis Libri XVIII,* Leyden, 1633.

*Delanglez, J., "Franquelin, Mapmaker," *Mid-America,* XXV, 1943, 29-74.

*——————., *Hennepin's Description of Louisiana,* Chicago, 1941.

*——————., "La Salle's Expedition of 1682," *Mid-America,* XXII, 1940, 3-37.

*——————., Marquette's Autograph Map of the Mississippi River, *Mid-America,* XXVIII, 1945, 30-53.

*——————., "The 1674 Account of the Discovery of the Mississippi," *Mid-America,* XXVI, 1944, 301-324.

*——————., *Some La Salle Journeys,* Chicago, 1938.

*——————., "The Sources of the Delisle Map of America, 1703," *Mid-America,* XXV, 1943, 275-298.

*Delisle, C., "Lettre de M. Delisle a M. Cassini sur l'embouchure de la riviere de Mississipi," *Journal des Sçavans,* May 10, 1700, 201-206.

Denucé, J., *Oud-Nederlandsche Kaartmakers in betrekking met Plantijn,* 2 vols., Antwerp and The Hague, 1912-1913.

*Díaz del Castillo, B., *Historia verdadera de la conquista de la Nueva España,* 3 vols., Mexico City, 1939.

Dickinson, R. E., and Howarth, O. J. R., *The Making of Geography,* Oxford, 1933.

*Dudley, R., *Dell' Arcano del Mare,* 3 vols. in 2, Florence, 1646-1647.

*Dunn, W. E., *Spanish and French Rivalry in the Gulf Region of the United States, 1678-1702. The Beginnings of Texas and Pensacola,* Austin, Tex., 1917.

*Duval, P., *Cartes de Geographie,* Paris, 1679.

*——————., *Le Monde; ov, La Geographie Vniverselle,* Paris, 1663, 1670.

——————., *Traité de Géographie qui donne la Connoissance et l'usage du globe et de la carte avec les Figures necessaires pour ce sujet,* Paris, 1672.

*Elvas, Gentleman of, *Relaçam verdadeira . . . feita per hũ fidalgo Deluas,* Evora, 1557.

Errera, C., *L'Epoca delle Grandi Scoperte Geografiche,* Milan, 1910º.

Favolius, H., *Theatri Orbis Terrarum Enchiridion,* Antwerp, 1585.

Feil, J., *Über das Leben and Wirken des Geographen Georg Matthaeus Vischer,* Vienna, 1857.

*Fernández de Navarrete, M., *Biblioteca Maritima Española,* 2 vols., Madrid, 1851.

*——————., *Colección de los Viajes y Descubrimientos que hicieron por mar los Españoles del Siglo XV,* 5 vols., Madrid, 1858-1880.

*——————., *Discertación sobre la Historia de Náutica, y de las Ciencias Matemáticas que han contribuido a sus Progressos entre los Españoles,* Madrid, 1816.

Final Report of the United States De Soto Expedition Commission, 76th Congress, 1st Session, House Document No. 71, Washington, D. C., 1939.

*Fischer, J., and von Wieser, R., *The Oldest Map with the name America of the year 1507 and the Carta Marina of the year 1516 by M. Waldseemüller (Ilacomilus),* Innsbruck, 1903.

Fite, E. D., and Freeman, A., *A Book of Old Maps Delineating American History from the Earliest Days Down to the Close of the Revolutionary War,* Cambridge, Mass., 1926.

Fontoura da Costa, A., "L'astronomie au Portugal à l'époque des grandes découvertes," *Comptes Rendus du Congrès International de Géographie Amsterdam, 1938,* 7 vols., Leyden, 1938, tome II, Travaux de la Section IV, Géographie Historique et Histoire de la Géographie, 13-24.

——————., "La lieue marine des Portugais au XVe et XVIe siècles," *ibid.,* 3-12.

——————., *A Marinharia dos Descobrimentos. Biografía Náutica Portuguesa atè 1700,* Lisbon, 1933.

Fordham, H. G., *Maps, Their History, Characteristics and Use,* Cambridge, 1927.

―――――――., *Some notable Surveyors & Map-makers of the sixteenth, seventeenth and eighteenth centuries, and their Work; a Study in the History of Cartography,* Cambridge, 1929.

―――――――., *Studies in Carto-bibliography,* Oxford, 1914.

Fournier, G., *Geographica Orbis Notitia per Litora Maris et Ripas Fluviorum,* Paris, 1667².

―――――――., *Hydrographie, contenant la Théorie et la Practique de toutes les Parties de la Navigation,* Paris, 1667².

*Frégault, G., *Iberville le Conquérant,* Montreal, °1944.

Frontières entre le Brésil et la Guyane Française. Mémoire présenté par les États-Unis du Brésil au gouvernement de la Confédération Suisse, arbitre choisi selon les stipulations du traité conclu à Rio-de-Janeiro, le 10 avril 1897 entre le Brésil et la France, 7 vols. in 6, Paris, 1899.― Vols. VI and VII form the "Atlas contenant un choix de cartes antérieures au traité conclu à Utrecht le 11 avril 1713 entre le Portugal et la France."

*―――――――., Second mémoire . . . entre le Brésil et la France, 6 vols., Paris, 1899. Volume VI is the atlas which is composed of two parts, "La première comprenant quatorze cartes antérieures au traité d'Utrecht, et un supplément à l'atlas annexé le 5 avril 1899. . . . La seconde partie renferme soixante-quinze cartes postérieures au traité d'Utrecht . . . accompagnées de notes."

*Gallois, L., *De Orontio Finaeo Gallico Geographo,* Paris, 1890.

―――――――., "L'Académie des Sciences et les origines de la carte de Cassini," *Annales de Géographie,* XVIII, 1909, 193-204, 289-310.

Gannet, H., "Certain Persistent Errors in Geography," *Bulletin* (formerly *Journal*) of the American Geographical Society of New York, XXXIII, 1901, 259-264.

*Garcilaso de la Vega, *La Florida del Ynca. Historia del Adelantado Hernando de Soto,* Lisbon, 1605.

*Garraghan, G. J., "Some Newly Discovered Marquette and La Salle Letters," *Archivum Historicum Societatis Jesu,* IV, 1935, 268-290.

*Gatschet, A. S., *A Migration Legend of the Creek Indians,* 2 vols., I, Philadelphia, 1884; II, St. Louis, 1888.

Gemma Phrysius, R., *De Principiis Astronomiae & Cosmographiae. Deqz vsu Globi ab eodem editi. Item de Orbis diuisione, & Insulis, rebusqz nuper inuentis,* Antwerp, 1530, Paris, 1557, Cologne, 1578.

*Gesellschaft für Erdkunde, ed., *Drei Karten von Gerhard Mercator. Europa―Britische Inseln―Weltkarte,* 3 parts in one volume, London, Berlin, Paris, 1891.

*Girava, H., *Dos Libros de Cosmographia,* Milan, 1556.

Graf, O., *Vom Begriff der Geographie im Verhältnis zu Geschichte und Naturwissenschaft,* Berlin, 1925.

Grande, S., *Le carte d'America di Giacomo Gastaldi,* Turin, 1905.

Griffin, A. P. C., *The Discovery of the Mississippi. A Bibliographical Account,* New York, 1883.

*Habig, M. A., *The Franciscan Père Marquette, A Critical Biography of Father Zénobe Membré, O.F.M.,* New York, °1934.

*Hakluyt, R., *The Principal Navigations Voyages Traffiques & Discoveries of the English Nation,* 12 vols., Glasgow, 1903-1905. Reprint of the second (1598-1600) edition.

Hamburgische Festschrift zur Erinnerung an die Entdeckung Amerika's, 2 vols., Hamburg, 1892.

Hamilton, R. N., "The Early Cartography of the Missouri Valley," *American Historical Review,* XXXIX, 1933-1934, 645-662.

Hamy, E.-T., *Etudes historiques et géographiques,* Paris, 1894.

*—————., *Note sur la Mappemonde de Diego Ribero (1529). Conservée au Musée de la Propagande de Rome,* Paris, 1887.

Hantzsch, V., *Sebastian Münster. Leben, Werk, Wissenschaftliche Bedeutung,* Leipzig, 1898.

*Hantzsch, V., and Schmidt, L., *Kartographische Denkmäler zur Entdeckungsgeschichte von Amerika, Asien, Australien und Afrika,* Leipzig, 1903.

*Harrisse, H., *Les Corte-Real et leurs voyages au Nouveau Monde,* Paris, 1883.

*—————., *The Discovery of North America,* Paris, 1892.

—————., *Découverte et Évolution Cartographique de Terre-Neuve et des pays circonvoisins 1497-1501-1769,* London and Paris, 1900.

*—————., *Jean et Sébastien Cabot. Leur origine et leurs voyages,* Paris, 1882.

—————., *John Cabot the discoverer of North America and Sebastian his Son,* London, 1896.

—————., *Notes pour servir à l'histoire, à la bibliographie et à la cartographie de la Nouvelle-France et des pays adjacents 1545-1700,* Paris, 1872.

Heawood, E., *A History of Geographical Discovery in the seventeenth and eighteenth centuries,* Cambridge, 1912.

*—————., *The Map of the World on Mercator's Projection by Jodocus Hondius Amsterdam 1608,* London, 1927.

Herrera y Tordesillas, A., *Description des Indes Occidentales, Qu'on appelle aujourdhuy le Novveau Monde,* Amsterdam, 1622.

*—————., *Historia General de los hechos de los Castellanos en las Islas i Tierra firme del Mar Oceano . . . En Cuatro Decadas desde el Ano de 1492 hasta el de 1531,* 2 vols., Madrid, 1601; the *Descripcion de las Indias ocidentales* is bound in with the second volume. The four subsequent decades were published in 2 volumes in Madrid, 1615. The 1730 and the 1934-1936 Madrid editions of Herrera's *Historia* begin with the *Descripcion de las Indias.*

—————., *Novvs Orbis, Sive Description Indiae Occidentalis,* Amsterdam, 1622.

Hessels, J. H., ed., *Ecclesiae Londino-Bataviae Archivvm.* Tomvs Primvs. *Abrahami Ortelli (Geographi Antverpiensis) et Virorvm ervditorvm ad evndem et ad Jacobvm Colivm Ortelianvm (Abrahami Ortelii sororis filivm) Epistvlae. Cvm Aliqvot aliis epistvlis et tractatibvs ab vtroqve collectis (1524-1628),* Cambridge, 1887.

*Heylin, P., *Cosmographie, The Fourth Book. Part II. Containing the Chorographie and Historie of America and all the Principal Kingdoms, Provinces, Seas, and Islands of it,* London, 1668.

Hildenbrand, F. J., *Matthias Quad und dessen Europae Universalis et Particularis Descriptio,* Frankenthal, 1890-1892.

*Hodge, F. W., ed., *Handbook of American Indians North of Mexico,* 2 parts, Washington, D. C., 1912[4].

*Hodge, F. W., and Lewis, T. H., eds., *Spanish Explorers of the Southern United States 1528-1543,* New York, 1907.

Holman, L. A., *Old Maps and Their Makers,* Boston, 1936[3].

Honter, J., *Rudimentorum cosmographicorum libri III,* Zurich, 1549, Antwerp, 1560, Zurich, 1578.

*[Huygen, J.], "Description de l'Amerique & des parties d'icelle, comme la Nouvelle France, Floride, des Antilles, Iucaya, Cuba Iamaica," bound in with *Le Grand Rovtier de Mer, De Iean Hvgves de Linschot Hollandois* . . . Nouvellement traduit de Flameng en François, Amsterdam, 1619.

*Jameson, J. F., ed., *Narratives of New Netherland 1609-1664,* New York, 1909.

Jardin, E., *Aperçu sur l'histoire de la cartographie d'après M. Charles P. Daly,* [Rochefort ? 1883?].

Jervis, W. W., *The World in Maps. A study in map evolution,* London, 1936.

*Jiménez de la Espada, M., ed., *Relaciones Geográficas de Indias,* 4 vols., Madrid, 1881-1897.

*Jomard, E. F., *Les monuments de la géographie,* Paris, 1842-1862.

*Karpinski Series of Reproductions. On these photographs of manuscript maps, cf. L. C. Karpinski, *Bibliography of the Printed Maps of Michigan, 1804-1880,* Lansing, Mich., 1931, 23-25.

Keane, J., *The Evolution of Geography,* London, 1899.

*Kellogg, L. P., ed., *Early Narratives of the Northwest 1634-1699,* New York, [e]1917.

Kenning, J., "Anschauungen über die Grösse der Erde im Zeitalter der Entdeckungen und ihre Beziehungen zu den älteren Gradmessungen," *Comptes Rendus* . . . *Amsterdam, 1938,* tome II, Travaux de la Section IV, 25-32.

Kimble, G. H. T., *Geography in the Middle Ages,* London, [1938].

Køppen, A. L., *The World in the Middle Ages: an Historical Geography,* New York, 1934.

*Kohl, J. G., *Die beiden ältesten General-Karten von Amerika. Ausgeführt in den Jahren 1527 und 1529 auf Befehl Kaiser Karl's V,* Weimar, 1860.

Kretschmer, K., "Die Atlanten des Battista Agense," *Zeitschrift* der Gesellschaft für Erdkunde zu Berlin, XXXI, 1896, 362-368.

*————., *Die Entdeckung Amerika's in ihrer Bedeutung für die Geschichte des Weltbildes,* text and atlas, Berlin, 1892.

————., *Geschichte der Geographie,* Berlin, 1912.

*Kunstmann, F., von Spruner, K., and Thomas, G. M., eds., *Atlas zur Entdeckungsgeschichte Amerikas,* Munich, 1859.

Labberton, R. H., *An Historical Atlas containing a chronological Series of one hundred Maps, at successive Periods, from the dawn of History to the present day,* Philadelphia, 1872.

Lafréry, A., publisher, *Tavole Moderne de Geografia,* [Rome, 1575?]

*Latorre, G., "La Cartografía Colonial Americana," *Bolétin* del Centro de Estudios Americanistas, Año III, 1915, n. 6, 1-10; n. 9 and n. 10, 1-14.

————., "Los geógrafos Españoles del Siglo XVI. Alonso de Santa Cruz," *ibid.*, Año I, 1913, n. 2, 29-51.

Lelewel, J., *Géographie du moyen age,* 5 vols., in 4, Brussels, 1852-1857. The atlas was published in 1850.

Lenglet Dufresnoy, N., *Methode pour étudier la Géographie,* 7 vols., Paris, 1741-1742[3].

*Leonard, I. A., "Don Andrés de Arriola and the Occupation of Pensacola Bay," *New Spain and the Anglo-American West,* 2 vols., privately printed, Los Angeles, 1932.

*————., *Don Carlos de Sigüenza y Góngora, A Mexican Savant of the Seventeenth Century,* Berkeley, Calif., 1929.

*————., *The Mercurio Volante of Don Carlos de Sigüenza y Góngora,* Los Angeles, 1932.

*————., *Spanish Approach to Pensacola, 1686-1693,* Albuquerque, N. Mex., 1939.

*————., "The Spanish Re-Exploration of the Gulf Coast in 1686," *Mississippi Valley Historical Review,* XXII, 1936, 547-557.

*Lewis, T. H., "The Chroniclers of De Soto's Expedition," *Publications* of the Mississippi Historical Society, VII, 1903, 379-387.

Leymarie, A. L., *Catalogue Illustré* (Exposition rétrospective des colonies françaises de l'Amérique du Nord), Paris, 1929.

Lloyd, F. E., "The delta of the Mississippi," *Journal of Geography,* III, 1904, 204-213.

*López de Gómara, F., *Primera y segunda parte de la historia general de las Indias,* Saragossa, 1553[2].

————., *Histoire Generalle des Indes Occidentales, . . .* traduite en François par le S. de Genillé Mart. Fumée, Paris, 1587.

*López de Velasco, J., *Geografía y Descripción Universal de las Indias,* Madrid, 1894.

*Lowery, W., *A Descriptive List of Maps of the Spanish Possessions within the Present Limits of the United States, 1502-1820,* Washington, D. C., 1912. The notes are by P. L. Phillips.

*―――――., *The Spanish Settlements within the Present Limits of the United States, 1513-1561,* New York, 1901.

―――――., *The Spanish Settlements within the Present Limits of the United States. Florida. 1562-1574,* New York, 1905.

*MacNutt, F. A., *Fernando Cortés his five letters of Relation to the Emperor Charles V,* 2 vols., Cleveland, 1908.

Marcel, G., *Cartographie de la Nouvelle France, Supplément à l'ouvrage de M. Harrisse,* Paris, 1885.

*―――――., *Reproductions de cartes et de globes relatifs à la découverte de l'Amérique du XVIᵉ au XVIIᵉ siècle,* avec le texte explicatif, Paris, 1892.

*Margry, P., ed., *Découvertes et Établissements des Français dans l'Ouest et dans le Sud de l'Amérique Septentrionale,* 6 vols., Paris, 1876-1888.

Marguet, F., *Histoire de la longitude à la mer au XVIIIᵉ siècle, en France,* Paris, 1917.

*Matal, J., *Speculum Orbis Terrae,* Cologne, 1600-1602.

*Mercator, G., *Atlas sive Cosmographicae Meditationes de Fabrica Mvndi et Fabricati Figvra,* Amsterdam, 1611⁴, 1630¹⁰.

*―――――., *Historia Mvndi: or Mercator's Atlas,* London, 1635, 1637.

*―――――., *Sphère terrestre et Sphère céleste de Gerard Mercator de Ruppelmonde éditées à Louvain en 1541 et 1551,* Brussels, 1875.

*Mercator, G., and Hondius, J., *Atlas Minor Gerardi Mercatoris à I. Hondio plurimis aeneis tabulis auctus et illustratus,* Amsterdam 1607, and 1634.

*Molinari, D. L., *El Nacimiento del Nuevo Mondo, 1492-1534, Historia y Cartografía,* Buenos Aires, ᶜ1941.

Moll, H., *Atlas Manuale: Or, A New Sett of Maps Of all the Parts of the Earth,* London, 1709.

―――――., *The World described: or, a New and Correct Sett of Maps,* [London, 1730?].

Morden, R., *Geography rectified,* London, 1688, 1700.

*Muller & Co., F., eds., *Remarkable Maps of the XVth, XVIth & XVIIth Centuries reproduced in their original size,* 6 parts in 4 vols., Amsterdam, 1894-1897.

*Münster, S., *Cosmographiae uniuersalis Lib. VI,* Basel, 1550, 1552; Italian translation, Cologne, 1575.

Murphy, H. C., *The Voyage of Verrazano: A Chapter of Maritime Discovery in America,* New York, 1875.

*Nordenskiöld, N. A. E., *Facsimile-Atlas to the early History of Cartography with reproductions of the most important maps printed in the XV and XVI centuries,* Stockholm, 1889.

*―――――., *Periplus, an Essay on the early History of Charts and Sailing-directions,* Stockholm, 1897.

Nunn, G. E., *The Geographical Conceptions of Columbus. A Critical Consideration of Four Problems,* New York, 1924.

―――――., *The Mappemonde of Juan de la Cosa. A critical Investigation of its Date,* Jenkintown, 1934.

*O'Callaghan, E. B., ed., *Documents Relative to the Colonial History of the State of New York,* vol. IX, Albany, 1855.
*O'Donnell, W. J., ed. and transl., "La Salle's Occupation of Texas," *Mid-America,* XVIII, 1936, 96-124.
*Ogilby, J., *America: being the latest and most accurate Description of the Nevv VVorld,* London, 1671.
*Ortelius, A., *Additamentum III Theatri Orbis Terrarum,* Antwerp, 1584.
*——————., *Theatrum Orbis Terrarum,* Antwerp, 1570, 1584.
*Oviedo y Valdés, G. F. de, *Historia General y Natural de las Indias, Islas y Tierra-firme del Mar Oceano,* 4 vols., Madrid, 1851-1855.
*Paullin, C. O., *Atlas of the Historical Geography of the United States,* New York, 1932.
Penne, C., Cassini, D., and Others, *Le Neptune françois, ou Atlas nouveau des Cartes Marines,* Paris, 1693.
*Phillips, P. L., *A List of Geographical Atlases in the Library of Congress,* 4 vols., I and II, Washington, D. C., 1909, III, 1914, IV, 1920.
*Pinart, A. L., *Recueil de Cartes, Plans et Vues relatifs aux États-Unis et au Canada, New York, Boston, Montréal, Québec, Louisbourg (1651-1731),* Paris, 1893.
"Points contestés et incertains dans l'interprétation des cartes, plus spécialement dans celles de l'époque des grandes découvertes," *Comptes Rendus . . . Amsterdam 1938,* tome II, Travaux de la Section IV, 146-205; papers by various authors.
*Priestley, H. I., ed., *The Luna Papers,* 2 vols., Deland, Fla., 1938.
*Ptolemy, C., *Geographiae Cl. Ptolemaei,* Rome, 1508.
*——————., *Geographiae opus,* Augsburg, 1513.
*Purchas, S., *Haklvytvs Posthumus or Pvrchas his Pilgrimes. Contayning a History of the World, in Sea voyages. & lande-Trauells, by Englishmen & others,* 4 vols., London, 1625.
Quad, M., *Compendium Vniversi Complectens Geographicarvm Enarrationvm Libros sex,* Cologne, 1600.
——————., *Fascicvlus Geographicvs Complectens Praecipvarvm Totivs orbis Regionum tabulas circiter centum. vnà cum earundem Enarrationibus,* Cologne, 1608.
——————., *Geographisch Handtbuch,* Cologne, 1600.
Ramusio, G. B., *Delle Navigationi et Viaggi,* 3 vols., Venice, I, 1554², II, 1559, III, 1556.
Recopilación de Leyes de Reynos de las Indias, 4 vols., Madrid, 1756².
*Richelet, P., *Histoire de la conquête de la Floride ou Relation de ce qui s'est passé dans la découverte de ce Pays par Ferdinand de Soto,* 2 vols., Paris, 1670.
*Richman, I. B., *California under Spain and Mexico 1535-1847,* Boston and New York, 1911.
*Robertson, J. A., ed. and transl., *Relaçam verdadeira . . . feita per hũ fidalgo Deluas,* Evora, 1557, 2 vols., I, text, II, translation, Deland, Fla., 1932-1933.

*Roland, F., *Alexis Hubert Jaillot, géographe du roi Louis XIV (1632-1712)*, Besançon, 1919.

——————., *Un franc-comtois éditeur et marchand d'estampes à Rome au XVIe siècle. Antoine Lafréry (1512-1577)*, Besançon, 1911.

Rollet de l'Isle, N., "Les Archives du Service Hydrographique de la Marine française," *Revue Hydrographique*, May 1924, 143-153.

Rosaccio, G., *Mondo Elementare et Celeste*, Trevigi, 1604.

Ruge, S., *Die Entwicklung der Kartographie von Amerika bis 1570*, in *Petermanns Mitteilungen*, XXIII, 1893, Ergänzungsheft no. 106.

——————., "The development of the Cartography of America up to the Year 1570," *Annual Report* of the Smithsonian Institution for 1894, Washington, D. C., 1896, 281-296.

Sandler, C., *Die Reformation der Kartographie um 1700*, text and atlas, Munich and Berlin, 1905.

*Sanson, N., *L'Ameriqve en plvsievrs cartes, & en divers traittés de geographie, et d'histoire*, Paris, 1657.

*——————, *Atlas Nouveau, Contenant Toutes les parties du Monde, ou Sont exactement Remarqués Les Empires, Monarchies, Royaumes, Estats, Republiques & Peuples qui sy trouuent à present*, Paris, 1692, and 1696; Amsterdam, 1700.

——————., *Introduction à la géographie*, Amsterdam, 1708.

*—————— and Sons, *Description de tout l'univers en plusieurs cartes*, 6 parts in 2 vols., Amsterdam, 1700.

*Santa Cruz, A. de, *Islario General de Todas las Islas del Mondo*, A. Bláquez, ed., text and atlas, Madrid, 1920.

——————., *Libros de las Longitudines*, A. Blásquez, ed., Seville, 1921.

*Scaife, W. B., *America, Its Geographical History, 1492-1892. . . . With a supplement entitled: Was the Rio del Espiritu Santo of the Spanish Geographers the Mississippi?* Baltimore, 1892.

——————., "Geographical Latitude," *Annual Report* of the Smithsonian Institution for 1889, Washington, D. C., 1890, 749-793.

Senex, J., *A New General Atlas . . . of the World*, 2 vols., London, 1721.

[Speed, J.], *A Prospect of the most Famous Parts of the World*, London, 1646, 1675, 1676.

*Sprengel, M. C., *Ueber Diego Riberos Welt-Karte von 1529*, Weimar, 1795.

*Steck, F. B., *The Jolliet-Marquette Expedition, 1673*, Quincy, Ill., 1928.

Stevens, H., *Historical and Geographical Notes, 1453-1530*, London and New Haven, Conn., 1869.

——————., *Johan Schöner, Professor of Mathematics at Nuremberg. A Reproduction of his Globe of 1523*, London, 1888.

Stevenson, E. L., *Atlas of Portolan Charts. Facsimiles of Manuscripts in the British Museum*, New York, 1911.

——————., *A Description of Early Maps and Facsimiles (1452-1611)*, New York, 1921.

——————., "Early Spanish Cartography of the New World, with special reference to the Wolfenbüttel-Spanish map and the work of Diego Ribero," *Proceedings* of the American Antiquarian Society, XIX, 1908-1909, 369-419.

*——————., "The Geographical Activities of the Casa de la Contratación," *Annals* of the Association of American Geographers, XVII, 1927, 39-59.

*——————., *Maps Illustrating Early Discovery and Exploration in America (1502-1530)*, New Brunswick, N. J., 1906.

*——————., *Marine World Chart of Nicolo Canerio Januensis. 1502 (circa). A Critical Study with Facsimile*, New York, 1908.

——————., "Martin Waldseemüller and the Early Lusitano-Germanic Cartography of the New World," *Bulletin* of the American Geographical Society, XXXVI, 1904, 193-215.

*——————., *Nova Universi Terrarum Orbis Mappa . . . duobus planispheris graphicè depicta à Guiliel Janssonio Alcmar*, New York, 1914.

——————, *Portolan Charts, their Origin and Characteristics*, New York, 1910.

——————., *Terrestrial and Celestial Globes. Their History and Construction*, 2 vols., New Haven, Conn., 1921.

——————.,*Willem Janszoon Blaeu 1571-1638. A Sketch of his Life and Work*, New York, 1914.

*—————— and Fischer, J., *Map of the World by Jodocus Hondius*, New York, 1907.

*Steward, J. F., "La Salle a Victim of His Error in Longitude," *Transactions* of the Illinois State Historical Society for the Year 1910, Springfield, Ill., 1912, 129-136.

*Stokes, I. N. P., *The Iconography of Manhattan Island, 1498-1909*, 6 vols., New York, 1915-1928.

*Swanton, J. R., and Halbert, H. S., *A Dictionary of the Choctaw Language by Cyrus Byington*, Smithsonian Institution Bureau of American Ethnology, *Bulletin 46*, Washington, D. C., 1915.

*Thevet, A., *La Cosmographie Universelle D'André Thevet Cosmographe du Roy*, Paris, 1575.

Tiele, P. A., *Nederlandsche Bibliographie van Land- en Volkenkunde*, Amsterdam, 1884.

*Thwaites, R. G., ed., *The Jesuit Relations and Allied Documents*, 73 vols., Cleveland, 1896-1901.

*Tucker, S. J., *Indian Villages of the Illinois Country*, vol. II, Scientific Papers, Illinois State Museum, Part I, *Atlas*, Springfield, Ill., 1942.

Uzielli, G., and Amat di San Filippo, P., *Mappamondi, Carte nautiche, Portolani ed altri Monumenti specialmente italiani dei secoli XIII-XVII*, Rome, 1882; vol. II of *Studi biografici e bibliografici* of Amat di San Filippo.

Vallaux, C., *Les sciences géographiques*, Paris, 1925.

Vallée, L., "Notice sur les documents exposés à la section des Cartes, Plans et Collections Géographiques du départment des imprimés de la Bibliothèque Nationale," *Revue des Bibliothèques,* XXII, 1912, 137-185.

*Van Ortroy, F., *L'oeuvre cartographique de Gérard et de Corneille de Jode,* Ghent, 1914.

——————., *Bibliographie de l'oeuvre de Pierre Apian (1495-1552),* Paris, 1901.

Van Raemdonck, J., *Gérard Mercator. Sa vie et ses oeuvres,* St. Nicolas, 1869.

*Veitia Linaje, J. de, *Norte de la Contratación de las Indias Occidentales,* 2 vols., Seville, 1672.

*Villiers du Terrage, M. de, *L'expédition de Cavelier de la Salle dans le Golfe du Mexicque (1684-1687),* Paris, 1931.

Vivien de Saint-Martin, L., *Histoire de la géographie et des découvertes géographiques depuis les temps les plus reculés jusqu'á nos jours,* text and atlas, Paris, 1873-1874.

——————., and Rousselet, P., eds., *Nouveau dictionnaire de géographie universelle,* 7 vols., Paris, 1879-1894, and a supplement in 2 vols., Paris, 1895-1899.

*Wagner, H. R., *Cartography of the Northwest Coast of America to the Year 1800,* 2 vols., Berkeley, Calif., 1937.

*——————., *The Manuscript Atlases of Battista Agnese,* a reprint for private circulation from *The Papers* of the Bibliographical Society of America, XXV, 1931.

——————., *The Spanish Southwest 1542-1794. An Annotated Bibliography,* Albuquerque, N. Mex., 1937.

Wagner, H., "Gerhard Mercator und die ersten Loxodromen auf Karten," *Annalen* der Hydrographie und Maritimen Meteorologie. *Zeitschrift* für Seefart- und Meereskunde, XLIII, 1915, 299-311.

——————., "Zur Geschichte der Seemeile," *ibid.,* XLI, 1913, 393-413, 441-450.

Wauwermans, H., *Histoire de l'école cartographique belge et anversoise du XVI*ᵉ *siècle,* 2 vols., Brussels, 1895.

Weise, A. J., *The Discoveries of America to the year 1525,* New York, 1884.

*Wieder, F. C., *Monumenta Cartographica,* 5 vols., The Hague, 1925-1933.

*Wieser, F. von, *Die Karten von Amerika in dem Islario General des Alonso de Santa Cruz,* Innsbruck, 1908.

Wilmore, A., *The Groundwork of Modern Geography,* London, 1931[3].

*Winship, G. P., "The Coronado Expedition, 1540-1542," Fourteenth *Annual Report* of the United States Bureau of American Ethnology, 1892-1893, Washington, D. C., 1896, part I, 329-613.

*Winsor, J., *Cartier to Frontenac. Geographical Discovery of the Interior of North America in its Historical Relations 1534-1700,* Boston and New York, 1894.

162   *EL RIO DEL ESPÍRITU SANTO*

————., *The Kohl Collection (now in the Library of Congress) of Maps relating to America,* Washington, D. C., 1904.
————., *The Mississippi Basin. The Struggle in America between England and France 1697-1763,* Boston and New York, 1895.
*————., ed., *Narrative and Critical History of America,* 8 vols., Boston and New York, °1884-1889.
Wolkenhauer, W., "Aus der Geschichte der Kartographie. Das Reformzeitalter der Kartographie. Gerhard Mercator, Jacopo Gastaldi, Philip Apian, Abraham Ortelius," *Deutsche Geographische Blätter,* XXXIII, 1910, 239-364.
————., *Leitfaden zur Geschichte der Kartographie in tabellarischer Darstellung,* Beslau, 1895.
Wright, J. K., *Aids to Geographical Research, Bibliographies and Periodicals,* New York, 1923.
*Wytfliet, C., *Descriptionis Ptolemaicae Augmentum,* Louvain, 1597.

LIST OF MAPS

The following list does not include all the maps which were studied or consulted to write this essay. There was no point in listing those which only delineate the American Coast or which contain little or no nomenclature. Nor would anything have been gained by listing the printed maps issued by various publishers during the seventeenth century. The latter maps show no originality, the variations noted on them are no proof that the draughtsmen had a better knowledge of the regions represented on their maps, since most of them are simply copies, often pirated copies of maps put out by internationally known map publishers. After 1681, the year of publication of Thévenot's *Recueil,* no map based on earlier patterns has been entered.

The Roman numerals in parentheses refer to the group to which the map belongs. The figures at the end of each entry refer to the page and note where the map is mentioned or discussed in this essay.

[1502?]

The Cantino map.—11, note 1.

[1503?]

The Caneiro map.—11, note 2.

1507

Map of the World, by M. Waldseemüller.—11, note 3.

1508

The Ruysch map.—11, note 5.

1513

The Admiral's map.—11, note 4.

1516

Waldseemüller's Carta Marina.—11, note 3.

[1520?]

(I)—The "Pineda" map of the Gulf of Mexico.—12, note 7.

LIST OF MAPS

ca. 1523
(II)—The Turin World map.—34, note 1.

1524
(II)—The "Cortés" map of the Gulf of Mexico.—14, note 11.

after 1524
(II)—The Paris Gilt Globe.—34, note 2.

1525-1527
(I)—The "Salviati" map.—16, note 17.

1527
(I)—The Maggiolo map of the World.—15, note 14.

1527
(I)—The Weimar map.—20, note 24.

after 1527
(I)—The Wolfenbüttel map.—21, note 25.

1529
(I)—The Ribeiro-Weimar map.—21, note 26.

1529
(I)—The Ribeiro-Borgia map.—21, note 27.

[1529]
(I)—The Verrazano map.—15, note 15.

1531
(II)—Double-cordiform globe, by Finaeus.—35, note 35.

1536
(I)—The map of Alonso de Chaves.—24-25.

1538
(II)—Mercator's reproduction of Finaeus' globe.—35, note 35.

1541
(I)—Map of the World, by N. Desliens.—29, note 40.

1541
(I)—Terrestrial Sphere, by G. Mercator.—26-27.

1542
(II)—The Euphrosynus Ulpius globe.—35, note 5.

1542
(I)—Map of the Gulf of Mexico, in John Rotz' atlas.—31, note 42.

1542
(I)—Map of the World, by Alonso de Santa Cruz.—27, note 36.

after 1542
(I)—The Dieppe or Harleian map of the World.—31, note 44.

1544
(I)—Map of the World, by Sebastian Cabot.—28, note 39.

after 1544
(III)—The "De Soto" map of Florida.—61-64.

1546
(I)—The "Henry II" map, by Pierre Desceliers.—31, note 42.

1547
(I)—Map of New Spain, in Nicolas Vallard's atlas.—31, note 42.

LIST OF MAPS

1550
(I)—Map of the World, by Pierre Desceliers.—30, note 41.

1550
(II)—The Western Hemisphere, by Diego Gutiérrez.—42, note 22.

after 1550
(I)—Most of the maps showing the Gulf of Mexico in Agnese's manuscript atlases belong to the first group.—25, note 33.

after 1550
(II)—Map of New Spain, by Agnese.—3, note 2; 36, note 7.

after 1550
(I)—Map of the Gulf of Mexico, in a Portuguese atlas in the Riccardiana Library, Florence; reproduced in Kretschmer, *Die Entdeckung Amerika's,* Taf., XXV, described in text, p. 329.

1554
(II)—Map of the World attributed to Gastaldi.—36, note 8.

1554
(I)—Map of the World, by L. Homem.—3, note 7.

1556
(II)—The Western Hemisphere, by Vopel.—35, note 6.

1558
(I)—Map showing the west coasts of Europe and Africa and the east coasts of North and South America; it is signed "Bastiam Lopez a fez 1558 nouembro 15." The original is in the British Museum, Add. Mss., 27303; photograph in the Library of Congress.

1558
(I)—Map of the Gulf of Mexico, in D. Homem atlas.—British Museum, Add. Mss., 5451a. Cf. Cortesão, *Cartografia e cartógrafos,* I, 373. An earlier state of the same map is in the Bibliothèque Nationale (BN), Paris, 5086.

ca. 1560
(II)—Map of America, by F. Berteli.—37, note 10.

after 1560
(I)—Map of the Gulf of Mexico, in Santa Cruz' *Islario General.*—28, note 38.

after 1560
(I)—Map of the Gulf of Mexico, in the "Duchess of Berry" atlas.—Paris, BN, Res. Ge. FF. 14409.

after 1561
(I)—Anonymous map of the Gulf of Mexico.—6, note 13.

1562
(II?)—Map of America, by D. Gutiérrez.—41.

1562
(II)—Map of America, by J. Cock.—37, note 2; 41.

1564-1565
(II)—*Infra,* 1594.

1565
(II)—Reprint by De Jode of Gastaldi's map of 1546.—Cf. Van Ortroy, *L'oeuvre cartographique,* 3.

1565
(II)—Map of the World, by P. Veronese.—37, note 10.

1566
(II)—Map of North America, by Zaltieri.—37, note 11.

before 1567
(II)—Santa Cruz' map of the Gulf of Mexico from the description in López de Velasco.—5.

1568
(I)—Map of the Gulf of Mexico, in D. Homem's atlas.—Reproduced in Hantzsch and Schmidt, *Kartographische Denkmäler,* Taf. V.

1568
(II)—Gulf of Mexico and Antilles, in F. Vas Dourado's atlas, Palacio de Liria, Madrid.—38, note 13.

[1568?]
(II)—Gulf of Mexico and Central America, in F. Vaz Dourado's atlas, Biblioteca Nacional, Lisbon.—38, note 14.

1569
(II)—G. Mercator's planisphere.—39, note 17.

1570
(IIa)—Western Hemisphere, in Ortelius' *Theatrum.*—43, note 23.

1570
(I)—Carte cosmographique ou Universelle description du monde auec le vrai pourtraict des vens. Faict en Dieppe par Jehan Cossin, marinier, en l'an 1570.—Paris, BN, Ge. D 7896.

after 1570
(II)—Map of the Gulf of Mexico in Dourado's atlas; original in the Huntington Library, San Marino, Calif., HM, 41; photograph in the E. E. Ayer Collection of the Newberry Library, Chicago. Cf. Cortesão, *Cartografia e cartógrafos,* II, 59-64.

1571
(II)—New Spain, Gulf of Mexico and Antilles, in Dourado's atlas, Torre do Tombo, Lisbon.—38, note 15.

[1573?]
(I)—Gulf of Mexico and Antilles, in Dourado's atlas, British Museum.—38, note 16.

1575
(IIa)—The Western Hemisphere, in Thevet's *Cosmographie.*—43, note 24.

[1578?]
(II)—The Gulf of Mexico and the Caribbean Sea, in Martínez' atlas.—44, note 28.

1580
(I)—The Gulf of Mexico, in Dourado's Munich atlas.—38, note 16.

1580
(I)—Terra Florida, in J. Oliva's atlas.—6, note 12.

1580
(I)—Spherical projection of a part of the northern hemisphere including all America north of the Equator, and the western part of Europe and Africa, by John Dee.—British Museum, Cottonian Aug. 1, 1, art. 1; reproduced in Hakluyt, *Principal Navigations,* end of vol. VIII.

1582
(II)—North America from the Tropic of Cancer to the North Pole, the Atlantic Ocean and the West coasts of Europe and Africa, by Michael Lok.—Reproduced in Winsor, *Narrative and Critical History of America,* III, 40.

1584
(I)—Terres Neufues, la Floride, les Neufues Espaignes, le Perv, le Bresil.—Paris, BN, Ge. C 4052 (K). On this map cf. Vallée, "Notice sur les documents . . . ," in *Revue des Bibliothèques,* XXII, 1912, 169.

1584
(III)—Map of Florida, by H. Chaves.—71, note 46.

after 1584
(III)—Map of America, by Mazza.—77, note 50.

1587
(IIa)—Map of the World, by R. Mercator.—44, note 25.

1587
(II)—The Gulf of Mexico on the map of the Northern Hemisphere, by C. Sgrooten.—44, note 28.

1587
(II)—Two maps of the Gulf of Mexico, in Martínez' atlas, Madrid. Bib. Nac., CC 35 ER 6, nos. 14 and 16.—44, note 28.

1589
(IIa)—Map of the World, by C. De Jode.—44, note 26.

1592
(I)—Map of the east coast of America from latitude 17° N. to latitude 37° N. "Thomas Hood made this platte 1592."—Reproduced in Kunstmann, *Atlas,* Blatt XIII.

1592
(III)—Map of the World, by Peter Plancius.—77, note 52.

1593
(IIa)—Northern Hemisphere, by G. De Jode.—45, note 29.

1593
(III)—North America, by C. De Jode.—77, note 53.

1594
(II)—Map of Florida, by Jacques Le Moyne. The Map was made in 1564-1565, but was first published by De Bry in his *Americae Pars Sexta,* Frankfurt, 1594.

1594
(IIIa)—Map of the World, by P. Plancius.—79, note 66.

1596
(IIIa)—The Western Hemisphere, in De Bry's *Americae Pars Sexta.*—79, note 67.

1597
(IIIa)—The "Christian Knight" planisphere, by J. Hondius.—79, note 68.

1597
(III)—Map of Florida, by C. Wytfliet.—78, note 54.

ca. 1598
(I)—Anonymous English map showing the east coast of America.—British Museum, Mss. 17938; reproduced in *Atlas brésilien,* no. 44.

1599
(IIIa)—The Plancius-Vrient mappamundi.—79, note 69.

1599
(I)—First state of a world map attributed to Ed. Wright.—Reproduced in *Atlas brésilien,* no. 49. A later state of the same map is reproduced in Nordenskiöld's *Facsimile-Atlas,* pl. L, and is attributed to R. Hakluyt.

17th cent.
(I)—Map of the eastern coast of North America.—British Museum, Add. Mss., 31858; photograph in the Library of Congress. Cf. Lowery, *A Descriptive List of Maps,* 103-104, t. 91.

17th cent.
(IIIa)—Map of America.—On this map, see Lowery, 102-103, t. 90.

17th cent.
(I)—Map of the Atlantic Ocean.—Paris, ASH, 116-17; partial reproduction in the Karpinski Series.

17th cent.
(I)—Map of the Gulf of Mexico, folio 24 of an anonymous manuscript atlas (no. 26) in the E. E. Ayer Collection of the Newberry Library, Chicago. Portuguese origin.

17th cent.
(I)—Anonymous map of the Atlantic Ocean.—Paris, BN, Ge. B 1148. Portuguese origin.

1600
(III)—New Spain, Florida and Mexico, by M. Tatton.—78, note 56.

after 1600
(III)—Map of the World, by F. Hoeius.—79, note 63.

1601
(I)—Map of the Atlantic Ocean, by G. Levasseur.—Paris, ASH, 116-6; photograph in the Karpinski Series of Reproduction.

1602
(III)—Map of Florida, by J. Matal.—78, note 55.

1605
(III)—Map of the World, by W. Blaeu.—78, note 57.

1607
(IIIa)—The Western Hemisphere, in Mercator-Hondius' *Atlas minor.*—79, note 70.

1608
(IIIa)—Planisphere, in M. Quadus' *Fasciculus Geographicus*.—Reproduced in Winship, *The Coronado Expedition*, between pp. 408-409.

1608
(III)—Map of the World, by J. Hondius.—78, note 59.

1611
(IIa)—Western Hemisphere, by M. Mercator.—44, note 27.

1611
(III)—Map of the World, by J. Hondius.—78, note 59.

1611
(III)—Western Hemisphere, in Mercator's *Atlas major*.—79, note 60.

1613
(I)—Map of the Atlantic Ocean, by P. Devaux.—Paris, ASH, 116-7; photograph in the Karpinski Series of Reproductions.

1614
(III)—The Western Hemisphere, published by P. Van den Keere.—79, note 61.

1620
(III)—Map of the Western Hemisphere, by M. Merian.—79, note 62.

1621
(IIIa)—Globe in twelve gores, by P. Goos.—79, note 71.

[1625?]
(IIa)—Novvelle Description Hydrographicque de TovT Le Monde, by N. Guerard.—Paris, ASH, 1-2; photograph in the Karpinski Series of Reproductions.

1625
(I)—The east coast of the Western Hemisphere, by J. Dupont.—Paris, ASH, 116-8-2; photograph in the Karpinski Series of Reproductions.

1625
(I)—North America, by H. Briggs.—Facsimile in Nordenskiöld's *Periplus*, pl. LX.

1625
(IV)—Map of Florida, in J. De Laet's *Nieuwe Wereldt*.—7, 81; note 1, 82.

1626
(III)—Map of the Western Hemisphere, engraved by P. Goos.

1630
(I)—Map of the Gulf of Mexico,, in R. Dudley's *Arcano del Mare*.—This map is slightly different in the 1646-1647 edition. The nomenclature on both maps is an indiscriminate compilation of the nomenclatures on maps which were in print at the time.

1631
(I)—Map of the Gulf of Mexico inserted in the Pas Caerte van Guinea, Brasilien en West Indien, by J. Aertsz Colom.—Paris, ASH, 116-9; photograph in the Karpinski Series of Reproductions.

LIST OF MAPS

1631
(I)—North America, by J. Guerard.—Paris, ASH, 116-10. Cf. Lowery, *A Descriptive List of Maps*, 129-130, t. 114.

1634
(I)—Carte Vniverselle Hydrographiqve, by J. Guerard.—Paris, ASH, 1-0-3; photograph in the Karpinski Series of Reproductions.

1635
(III)—The Western Hemisphere, by W. Janszoon Blaeu.—78, note 58. From this date on, this map is reproduced in the various atlases issued by the Blaeu firm as well as in many compilations by other map publishers.

1639
(IV)—Map of North America, by Hondius.—7, 86, note 15.

[1640]
(IIIa)—The two Hemispheres, by D. Danckerts.—79, note 72.

ca. 1640
(III)—The Western Hemisphere, by C. De Jonghe.—Reproduced in *Atlas brésilien*, no. 65.

1642
(I)—The Gulf of Mexico and the Caribbean, by Jan Janszoon.—This map is reproduced in every atlas published by the Janszoon firm. On this map publisher see 86, note 16.

1648
(III)—Map of the World, by J. Blaeu.—Reproduced in Wieder, *Monumenta cartographica*, III.

1650
(IV)—Map of North America, by N. Sanson.—7, 82, note 8, 87.

1653
(IV)—Map of North America, by J. Janszoon.—79, note 15.

1656 & 1657
(III and IV)—New Mexico and Florida, by N. Sanson.—88-91.

1660
(III)—Map of the Western Hemisphere, by F. De Wit.—Reproduced in Karpinski, *Bibliography of the Printed Maps of Michigan*, facing p. 90.

1663
(III)—Map of the Western Hemisphere, by P. Heylin.—79, note 64.

1663
(III and IV)—Map of Florida, by P. Duval.—110, note 38.

after 1663
(I)—Les Costes de l'Amerique dite autrement Nouueau Monde et Indes Occidentals, anonymous.—Paris, ASH, 117-0-26; photograph in the Karpinski Series of Reproductions.

1667
(I)—Anonymous map of the Gulf of Mexico, Central America and the West Indies.—Paris, ASH, 140-1-1; photograph in the Karpinski Series of Reproductions.

1669
(III and IV)—Map of North America, by G. Sanson.—93, note 24.

1671
(III)—Map of the Western Hemisphere, by J. Ogilby.—79, note 65.
(I)—Nova Hispania Nova Galicia Gvatimala in Ogilby's *America*.

1671
(I)—Map of the Gulf of Mexico and the Caribbean, by A. Montanus, in his *De Nieuwe en Onbekende Weereld*.—This map is merely a variant of that listed above under 1642.

1672
(I)—The Gulf of Mexico and the West Indies; by H. Doncker, in his *Atlas de la Mer, ou, Monde Aquaticque*.

1674
The Sanson-Jaillot map of North America.—92, note 23.

1674
Nouuelle Decouuerte de Plusieurs Natons Dans la Nouuelle France En L'annee 1673 et 1674; copy by an anonymous author of Jolliet's lost map.—Facsimile in colors in Thwaites' *Jesuit Relations*, vol. 59, facing p. 86.

after 1674
Carte de L'Amerique Septentrionale Depuis l'embouchûre de la Riviere St. Laurens jusques au Sein Mexique, by H. Randin.—Reproduced in Tucker, *Indian Villages of the Illinois Country*, pl. VI. Randin's map of America which he used as a basis has a "B. du S. Esprit" in the northwest corner of the Gulf, he has the Mississippi emptying into it.

1675
(I)—The Gulf of Mexico in Roggeveen's *Burning Fen*.—Reproduced in Winsor, *The Mississippi Basin*, 39.

[1678]
Carte Gnlle de la France Septen-Trionalle, by J. B. L. Franquelin.—Paris, Bibliothèque du Service Hydrographique (SHB), B 4040-11; photograph in the Karpinski Series of Reproductions. This map is a variant of that listed above under 1674.

1679
(III and IV)—North and Central America, by P. Duval.—94, note 25.

1681
Sectional map of North America from latitude 27° to latitude 44°, by J. B. L. Franquelin.—Paris, SHB, B 4040-4; photograph in the Karpinski Series of Reproductions. The basic map on which Franquelin inscribed the Mississippi River was similar to that used by Randin, *supra*, after 1674. Cf. text, p. 110.

1681
Carte de la decouverte faite l'an 1673 dans l'Amerique Septentrionale, in M. Thévenot's *Recueil de voyages.*—123.

[1682]
Carte de l'Amerique Septentrionale et partie de la Meridionale, by Bernou-Peronel.—110, note 36.

1683
Carte de la Nouuelle France et de la Louisiane, in L. Hennepin's *Description de la Louisiane.*—123.

1683
The Marly Globe, by M. V. Coronelli.—111.

1684
Carte de la Louisiane ou des voyages du Sr de la Salle, by J. B. L. Franquelin; reproduced in colors in Thwaites' *Jesuit Relations,* vol. 63, frontispiece—104, 107.

1685
Map of North America, by Minet.—108, note 31.

1686
Sketch map, by M. Echagaray.—121, note 14.

1687
The gores of the Florence Globe, by M. V. Coronelli; original in the Library of Congress, photograph in the E. E. Ayer Collection of the Newberry Library, Chicago.—111.

1688
America Settentrionale, by M. V. Coronelli in his *Atlante Veneto,* reproduced in Tucker, *Indian Villages of the Illinois Country,* plates IX and X.—111.

1693
Nueva Demarcación de la Bahía de S^a Maria de Galve (antes Pensacola), by C. Sigüenza y Góngora.—131, note 26.

after 1693
Anonymous sketch map of the Gulf Coast, the Mississippi River and the St. Lawrence.—139, note 22.

1696
Three drafts of the map of North America, by C. Delisle.—9, note 16; 136.

1699
Map of the Gulf Coast and of the lower course of the Mississippi River, by N. De Fer.—9, note 18.

[1699?]
Anonymous map of the northern Gulf Coast and of the Mississippi from the Illinois River to the sea.—144, note 43.

[1699]
Three maps of a section of the northern Gulf Coast, by Ste. Marie.—141, note 32.

1700

Globe terrestre, by Guillaume, *i.e.*, Claude Delisle, reproduced in *Atlas brésilien*, no. 87.—137.

1700

L'Amérique Septentrionale, by Guillaume, *i.e.*, Claude Delisle.—9, note 16; 137.

1701

Les Costes aux Environs de la Riviere de Misisipi, by N. De Fer.—9, note 18.

after 1701

Drafts of the map of 1703, by Claude Delisle.—Paris, Archives Nationales (AN), JJ, 178-1-14,15; 193-6,12; SHB, C 4040-4.

1703

Lower half of the map of North America, by Guillaume, *i.e.*, Claude Delisle. —9, note 17; 142.

1718

Carte de la Louisiane et du Cours du Mississipi, by Guillaume Delisle; the draft of this map is reproduced in Tucker, *Indian Villages of the Illinois Country*, pl. XI.—142.

Plate 1.

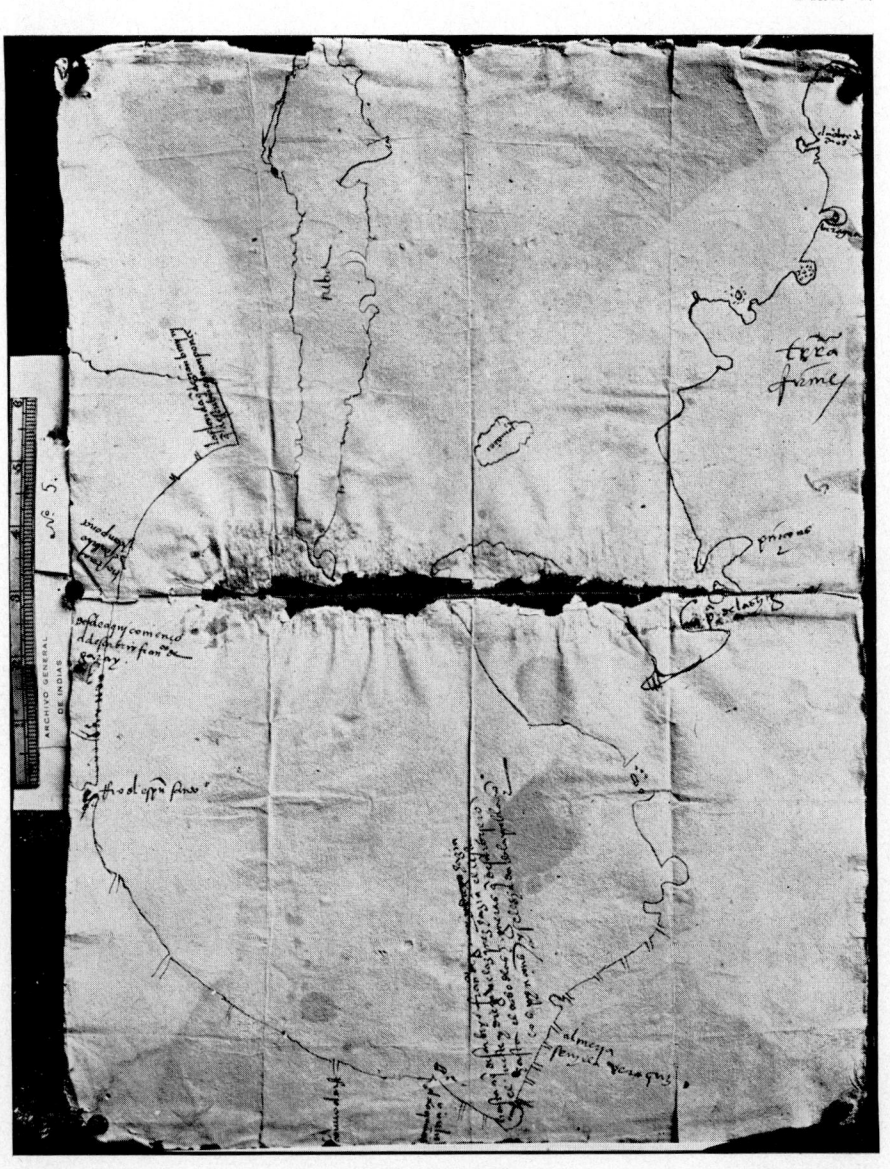

The "Pineda" Map.

(p. 12, note 7)

Plate 2.

THE GULF OF MEXICO ON THE RIBERO-WEIMAR MAP. (I)

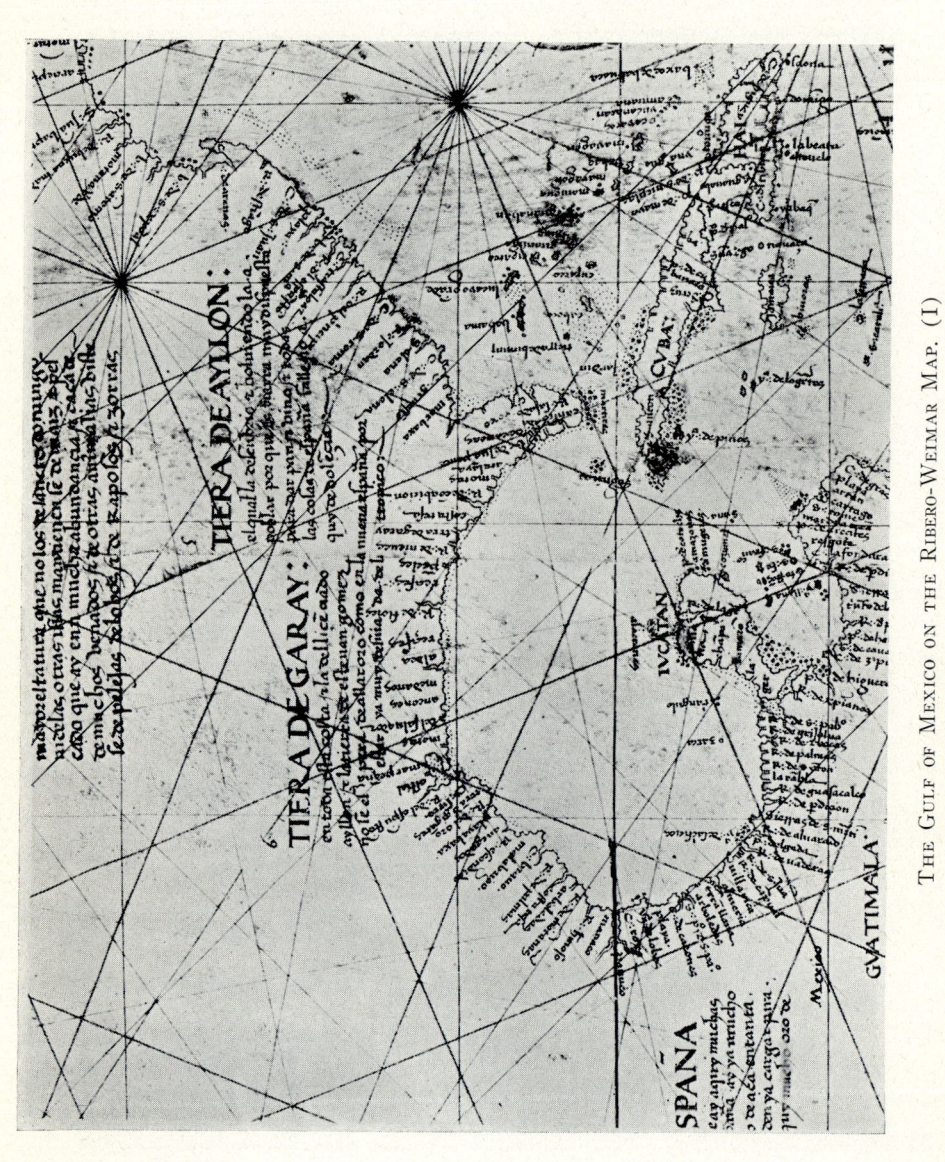

(p. 20, note 14)

Plate 3.

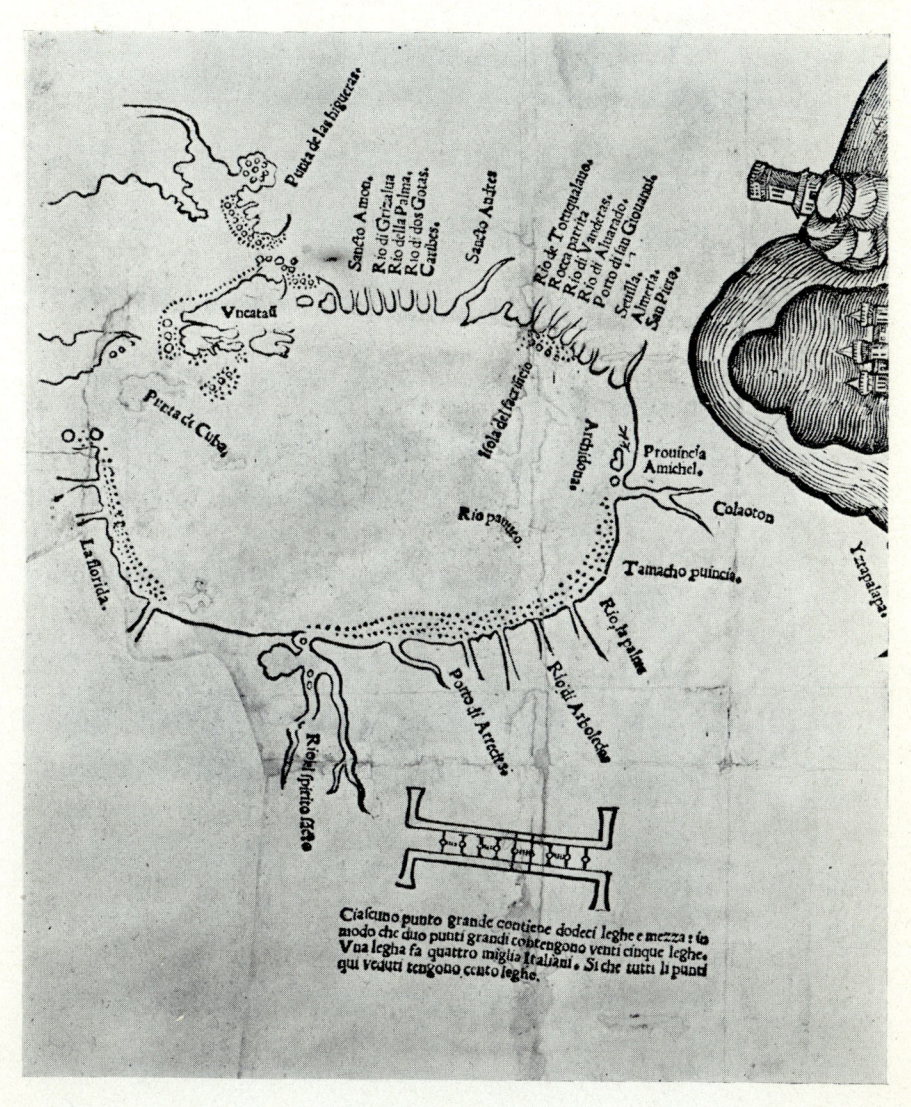

The "Cortés" Map. (II)

(p. 14, note 11)

Plate 4.

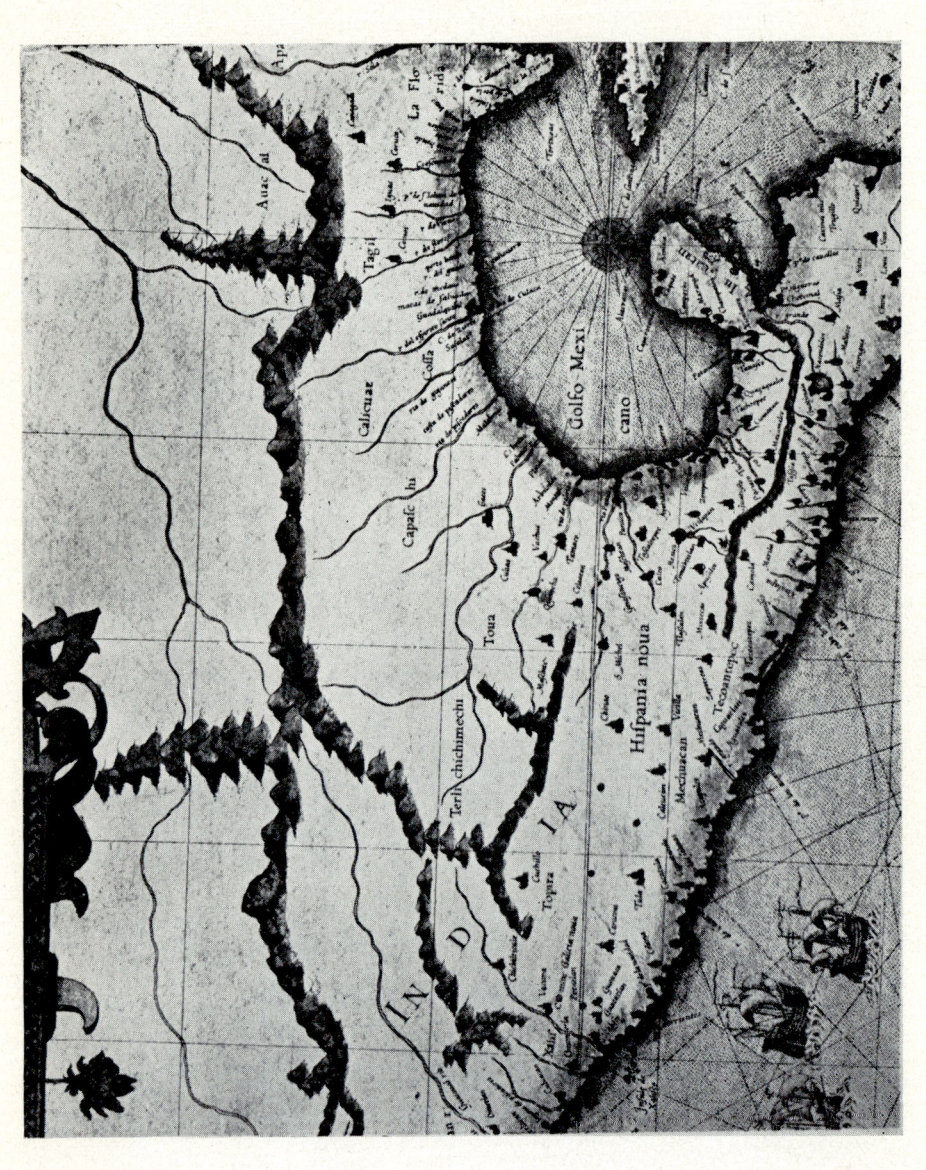

THE GULF OF MEXICO ON MERCATOR'S PLANISPHERE OF 1569. (II)

(p. 39, note 17)

Plate 5.

The Chaves Map of Florida. (III)

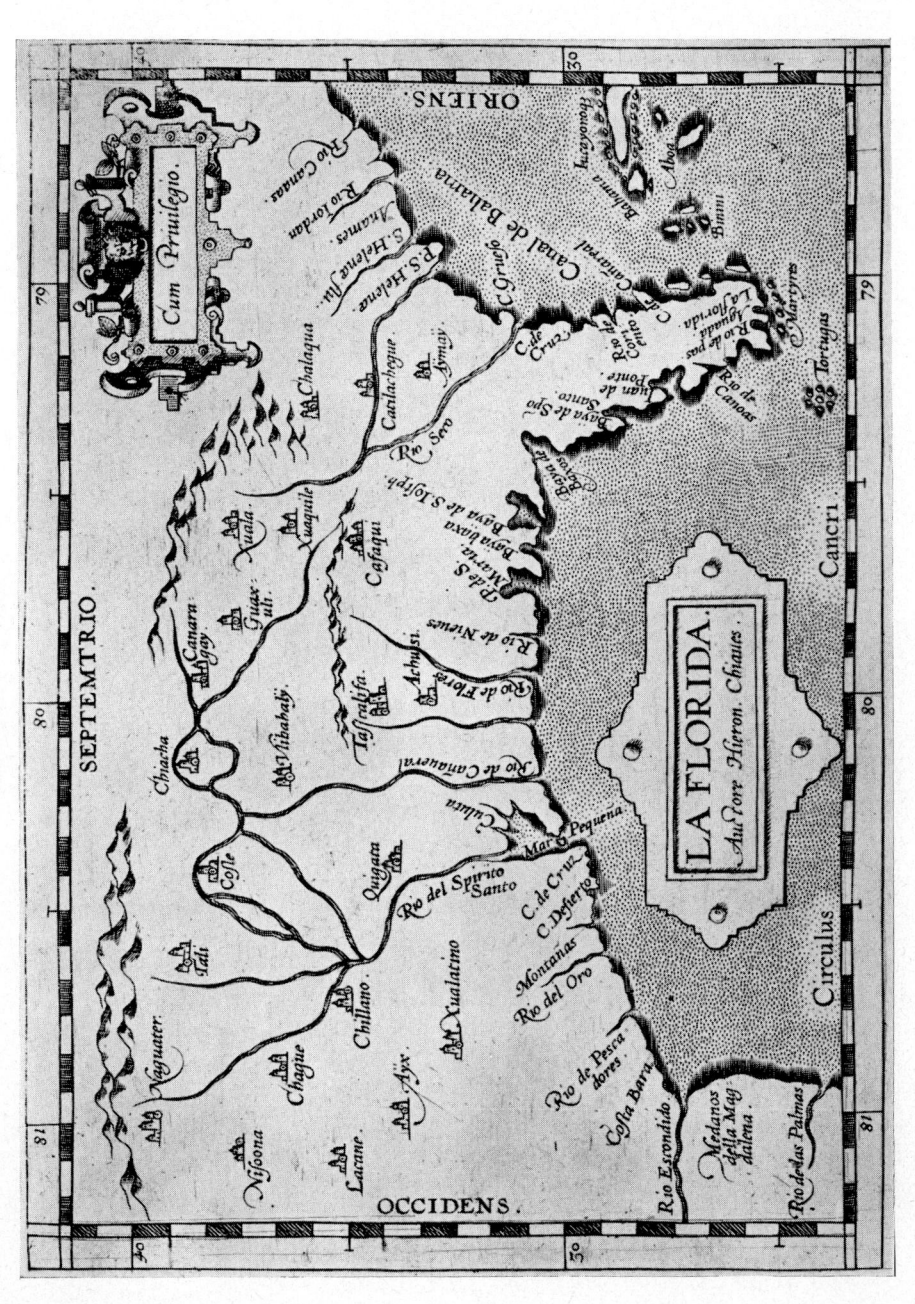

(p. 71, note 46)

Plate 6.

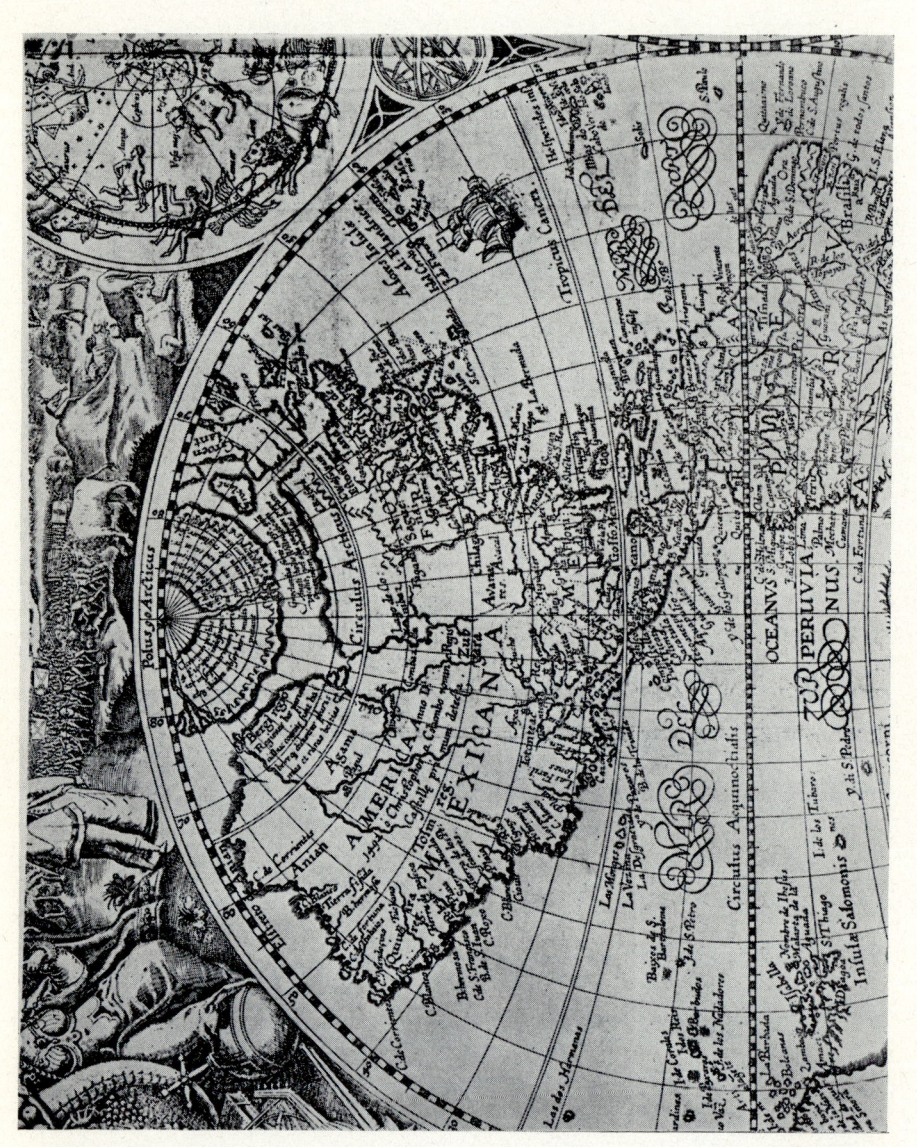

The Western Hemisphere on Plancius' Globe, 1594. (IIIa)

(p. 79, note 66)

Plate 7.

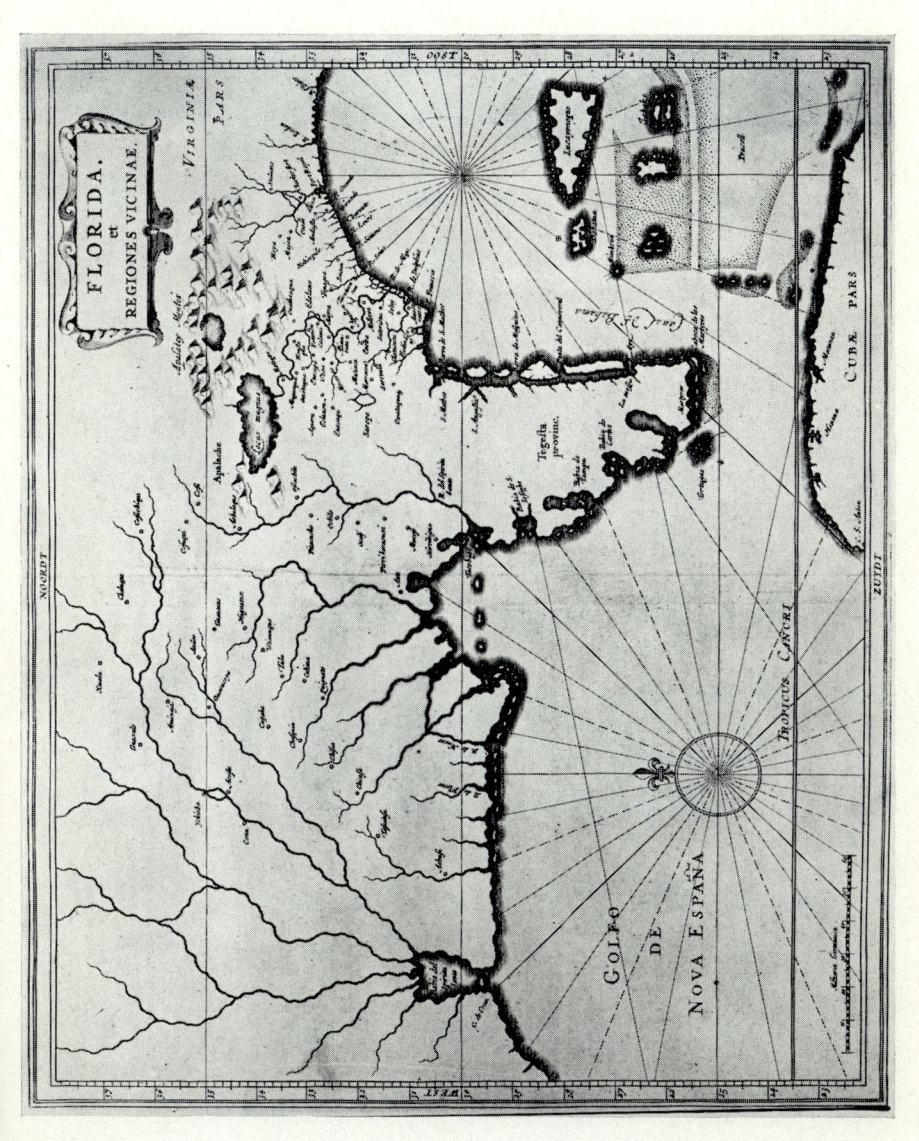

THE DE LAET MAP OF FLORIDA, 1625. (IV)

(p. 81, note 1)

Plate 8.

THE GULF OF MEXICO ON THE SANSON-JAILLOT MAP, 1674.

(p. 92, note 23)

INDEX

ABBREVIATIONS
B.—Bahía, Baie, Bay.
BES.—Bahía del Espíritu Santo.
C.—Cabo, Cap, Cape.
I.—Island.
Is.—Islands.
L.—Lac, Lago, Lake.
MR.—Mississippi R.
R.—Rio, River, Rivière.
RES.—Rio del Espíritu Santo.

A

Achalaque, 89
Achuse, Achusi, Achussi, Ochus, 58, 66, 69, 72, 73, 130; see Pensacola B.
Acoste, Coste, 85, 89.
Admiral's map, the, 11
Agile, Axile, 41
Agnese, G. B., maps, 2, 25, note 33, 36
Alabama, 11, 60
Alabama R., xii, 114
Alibano, 63, 92
Allard, C., map, 33
Alleghanies, 39
Allouez, J. C., 98, 99
Almirante, R. del, (Blackwater R.), 132, 135
Amichel, 13, 34
Aminoja, Aminoya, 85
Angeles, Angelos, R. de los, 26, 27, 30, 63, 75
Apalache, Apalchen, 40, 58, 115
Apalache, country, 51, 94; mission, 114; mountains, 93; port (St. Marks, Fla.), 113, 118, 119
Apalachee B., 8, 15, 22, 29, 34, 37, 47, 51, 64, 75, 83, 84, 86, 89, 90, 94, 108, 110, 118
Apalachee, C., 128
Apalachicola R., RES, 142
Apalchen, Apalache, 40
Appalachians, xii, 107
Arache, 70
Arboledas, R. de, 14, 34
Arkansas R., xi, xiii, 89, 102; village, 64

Arnoul, P., 107
Arrecifes, Arrecifos, Arrecife, punto de, 14, 15, 34
Arriola, A. de, 140; at the mouth of the MR, 134-135
Ascension, 32
Atlantic Coast, 75
Atlantic Ocean, North Sea, Sea of the North, 64, 70, 71, 100
Atlante Veneto, 111
Austin, Tex., 75, 76
Autiamque, Utiamgue, 89
Avacal, 40
Axile, Agile, 40
Ayataba, 63
Ayllón, L. Vásquez de, 85

B

Badine, the, 139, 140, 141, 143
Bandelier, A., 48
Barroto, J. Enríquez, 118, 119, 121, 125, 126
Barroto - Romero, first expedition, 120; second expedition, 123
Baudrand, M. A., 100
Baxo, Ancon, 84
Baxo, C., 25
Baxo, R., 75
Beaujeu, T. Le Gallois de, 107, 136, 138
Bernou, C., 135; writings on the MR, 109-111; on the mouth of the MR, 136
Bernou-Peronel, map, 110
Biedma, L. Hernández de, 40, 62, 72, 73; field notes, 64, 74; narrative of the De Soto expedition, 55, 57-61, 91; calls the MR Rio Grande and RES, 59, 61, 62; voyage of Maldonado, 57-58
Biloxi, B., 144
Blaeu, J., map, 7, 86-87
Blaeu, W., 86; maps, 6, 78
Blackwater R. (R. del Almirante), 132
Bon, the, 138

INDEX

Borgne, L., 140, 143
Bourne, E. G., 56; comments on Ranjel's narrative, 57
Bravo, C. (C. Brave), 26, 30
Bravo, R., 109, 110, 111, 120, 140
Breton Sound, identified as the BES, 137, 144
Burns I., 59

C

Caballos, B. de, 51
Cabeza de Vaca, A. Núñez, narrative, 47 ff; in Spain, 55
Cabot, S., 24, 29; map, 27, 28, 47, 49
Calcasieu Pass, La., 114
California, Gulf of, Mar Vermejo, Sea of California, Vermilion Sea, ix, 97, 99, 102
Calicuas, Calicuaz, Coligoa, Coligua, Colima, 40, 41, 88, 89
Cameron, La. 76
Campeche, 114, 115; B. of, 36
Campo, A. do, 71
Canada, 94, 96, 100, 109, 116
Canada R., the St. Lawrence R., 132
Canagali, 40
Canary Is., 114
Cañaveral, el, R. del, R. de Canaveral, 62, 74, 75, 76, 80, 90, 94, 101
Canerio, N. de, map, 11
Cantino, A., map, 11
Canoas, R. de, 25, 36
Cap Français, Cap Haitien, Haiti, 139
Capaha, 86
Capaschi, 40, 41
Capachequi, 41
Carlos B., Escampaba, 5, 127, 128
Casa de Contratación, passim 17-75, 114, 121; maps in the, 60, 123, 127
Cassini, J. D., 135
Castañeda, P. de, narrative of the Coronado expedition, 69-71
Castilla del Oro, 24
Castillo, A. del, 47, 48
Cathay (China), 34
Catilachegue, 72
Catoche, C., 128

Chaguet, 63
Charles II, 128
Charles V, 1, 19, 20, 21, 29, 55
Chasteaumorant, the marquis de, 140
Chattanooga, 60
Chattaux, R. des, Apalachicola R., 142
Chaves, A. de, 24, 29; maps, 24, 56; map of 1536, 2, 7, 22, 25, 26, 27, 28, 31, 57, 61, 62, 73
Chaves, H. de, 71-72, 73; map of Florida, 6, 88, 89; map of Florida compared with the De Soto map, hydrography, nomenclature, positions of towns, 71-78
Chiacha, Chiaha, Chiha, Ychiaha, 59, 89
Chicago, xi, 34, 35
Chicago R., 122
Chicasa, 63
Chickasaw Co., Miss., 60
Chickasaw villages, 60
Chico, puerto, 31
Chiha, see Chiacha
Chillano, 72, 73
China (Cathay), 34
Choctaw, 67
Chucagua, Choucagoua, Choucagua, 8, 70, 103, 105, 138; on Duval's map of 1679, 94-95; in Garcilaso de la Vega's narrative, 106; on Minet's map, 108; location, 106; is the MR, 85, 92, 94, 138; is not the MR, 104
Chuse, B., 73; see Achuse
Clairambault, P., 108
Cleburn Co., Ala., 90
Coahuila, 128, 130
Coça, Cossa, Coza, 40, 60, 63, 66, 85; site of village, 69
Coças, Creek Indians, 67, 68
Cock, J., map, 37, 39, 88; description, 41-43, source of Mercator's planisphere, 40-41
Cofitacheque, 72
Colaoton, 41
Colbert, R., the MR, 105, 11, 132; the R. de la Palizada, 134

INDEX

Coles, J., 92
Colima, 89, see Calicuas
Columbus, C., 17
Columbus, F., 21, 22
Comos, 40
Consulado de la Casa de Contratación, 114, 115
Coosa R., xii, 60, 66, 69
Coronado, F. Vásquez de, expedition, 69-71
Coronelli, M. V., 135; maps and globes, 8, 102, 111, 136
Corpus Christi Pass, 74
Cortés, H., map, 14, 15, 16, 33, 45; compared with the Turin map, 34; second letter, 1, 14
Coruco, 40
Cossa, 40, see Coça
Coste, Acoste, 89
Council of the Indies, 18, 115, 126, 127, 130
Courcelle, D. de Rémy de, 99
Coza, 40, see Coça
Creek Indians, Coças, 67
Cruz, B. de, 51; C. de la, 2
Cuba, 58, 61
Culata, 2, 4, 43; is the BES, 84
Culiacán, 70
Cutifachiqui, 72

D

Dablon, C., 98; describes the MR; account of the discovery of the MR, 101; MR probably RES, 109
Danckert, D., map, 79
Dávila Padilla, A., narrative, 66-69
Dawson, S. E., 11
De Bry, T., map, 79
De Fer, N., map, 9
De Graff, L., 140
De Jode, C. and G., maps, 44, 45, 77
De Laet, J., map of Florida, 7, 77, 81, 88, 89, 93, 98; hydrography and positions of towns, 85-86; nomenclature of Florida, 91; *Nieuwe Wereldt*, 7, 81; *Nova Hispania*, map, 83

Delisle, C., letter to Cassini, 135-136; globe and maps, 3, note 5, 9, 136, 137, 143-144; route of the De Soto expedition, 91-92
Delisle, G., 9, 144; map, 142
Denver, Colo., 76
Desceliers, P., map, 30-31
Desierto, C., C. d Zierto, 27
Desliens, N., 44; map, 29-30
De Soto, H., 39, 51, 70, 72, 135; at Seville, 57; chroniclers and narratives of the expedition, ix, 40, 41, 42, 43, 55 ff, 82, 85, 86, 88, 91, 93, 103, 119, 134; landing place, 3, 61, 83, 144; discovers RES, 69; route, 60, 86; route plotted by Delisle, 91-92
"De Soto" map, the, 6, 31, 40, 60-61, 65, 73; authorship, date, description, 61-64; compared with Chaves' map of Florida, 72-76
Des Plaines R., xi, 122
Dieppe, mapmakers of, 29-30
Dollier de Casson, F., 99
Dorantes, A., 47, 55; goes in search of R. Pánuco, 53-54
Dourado, F. Vaz, maps, 37-39
Dunn, W. E., 113, 116, 117, 118, 119, 120, 122, 124
Duval, P., map, 8, 94-95, 110

E

Echagaray, M., 116, 126; memorial, 113-115; map, 121-123
Elmore Co., Ala, 66
Elvas, the Gentleman of, 40, 41, 70, 72, 73, 85; calls the MR Rio Grande, 59; narrative of the De Soto expedition, 55, 56, 57, 91, 93, 103; voyage of Maldonado, 58-59
Empalizada, R. de la, 120, 121; see Rio de la Palizada
Erie, L., 93
Escampaba (Carlos B.), 5
Escondido, el Condido, C., 27, 32, 87
Escondido, R., 28, 62, 74, 106, 110; the MR, 104, 105, 111, 136, 137

INDEX

Española, I., 56
Espíritu Santo, B. del, 1-10, 25, 65, 103, 104, 106, 107, 108, 109, 113, 132; on the west coast of the Florida peninsula, 5-6, 84; Tampa Bay, 144; the two BES, 4-5; landing place of De Soto, 83, 144; La Salle's landing place and colony on the, 117, 120, 121, 123; identified as Breton Sound, 137, 144; as Culata, 84; Galveston Bay (?), 132, 145; Lago de Lodo, 143, 144; Mar pequeña, 2-3; Matagorda Bay, 125-126, 130, 143; Mobile Bay, 123-126, 128-129, 130, 145; Lake Sabine (?), 145; in Cabeza de Vaca's narrative, 48; in Echagaray's memorial, 113, 114, 115; in the *Jesuit Relations,* 97-99; in La Salle's writings, 105-107; location, 54, 117, 119, 120, 121, 127, 128-129, 137; location on maps of the Gulf, 125, 143; on Agnese's map, 36; on the Bernou-Peronel map, 110, on Coronelli's maps, 102; on De Laet's map, 83, 85; on Delisle's globe and maps, 137, 143-144; on the De Soto map, 64; on Echagaray's map, 121-122; on Mercator's globe of 1541, 27; on Palacios' maps, 116; on Sanson's map of 1650, 101, of 1656, 90; the western BES is not found on maps made after 1700, 142-143
Espíritu Santo, Puerto del, 2
Espíritu Santo, Rio del, xiii, 27, 67, 69, 108, 113, 117; on the west coast of the Floridan peninsula, 6; description, 65; fordable, 66, 67; identified as the Apalachicola R., 142; as the Rio Grande, 134, 135; probably the MR, 101; the MR, 61, 108-109; the Mississippi-Missouri, 70; Oquechiton, 67, 68; the Trinity R. (?), 132, 145; the RES is not the MR, 146; Maldonado to, 58, 61; the RES in Biedma's narrative, 57-61, 62; in Cabeza de Vaca's narrative, 53; in Castañeda's narrative, 69-71; in Dávila Padilla's *Historia,* 66-69; the RES in the *Jesuit Relations,* 98; in La Salle's writings, 103-107; in the *Luna Papers,* 65-66; in the narratives of the De Soto expedition, 55-57; location, 32, 54; on maps of the second group, 45; on seventeenth-century maps, 107; on maps of the third group, 79-80; on Spanish maps, 120; on Agnese's map, 36; on the Bernou-Peronel map, 110; on the Cabot map, 28; on the Chaves map of 1536, 25, 57, 74; on H. Chaves' map, 76; on Cock's map, 37; on Coronelli's map, 102, 111; on the Cortés map, 33, 34; on De Jode's map, 45; on De Laet's map, 83, 84-85; on Delisle's maps, 137, 142, 143; on the De Soto map, 63, 64, 75; on the Dourado maps, 38-39; on Duval's map, 94; on Finaeus' globe, 35; on Gastaldi's map, 36; on Mercator's planisphere, 43; on Rumold Mercator's map, 44; on the Paris Gilt Globe, 34; on Sanson's map of 1650, 88, of 1656, 90, 101; on the Sanson-Jaillot map, 94; on Santa Cruz's map, 28; on Thevet's map, 44; on the Turin map, 34; on Ulpius' globe, 35
Evora, 56

F

Ferdinand V, 18
Finaeus, globe, 35
Fleurs, Coste des, 30; R. des, 84
Flores, Costa de, 27; R. de, 30, 63, 83
Florida, 16, 34, 40, 45, 59, 60, 75, 97, 98, 100, 101, 103, 113, 115, 136; description 65, 85; see Southern United States
Florida, maps of 70, 123; Chaves' map, 72 ff; Sanson's map of 1650, 87, of 1656; 88-90; Sanson-Jaillot map, 93-94; nomenclature on De Laet's map, 91

INDEX

Florida Keys, 7, 9, 12, 130
Florida peninsula, 1, 5, 36, 83, 105, 107, 108, 127, 137, 144; west coast of, 4, 27, 105
Florida Sea, Gulf of Mexico, 98, 99
Fox, R., xi, 16, 98
François, the, 140
Franquelin, J. B. L., map of 1684, 8, 102, 107

G

Galinée, R. F. Bréhan de, 99, 100
Galve, Conde de, 127, 132; describes the Gulf of Mexico, 128-129
Galveston, B., 24, 25, 27, 43, 47, 52, 75, 76, 121, 124; the BES (?), 132, 145
Galveston, I., 52
Garay, F. de, 12, 15, 16
Garcilaso de la Vega, 70, 82, 85, 89, 104, 108; narrative of the De Soto expedition, 55, 86, 91-92, 103, 105, 106
Garcitas, R., 115, 125
Gastaldi, J., map, 36
Gatschet, A. S., 66
Georgia, 11
Gigantes, R. de, Tierra de, 27
Girava, H., 35
Goos, P., gores, 79
Graham Co., Kansas, 76
Grande, R., 89, 90, 94, 95, 103, 107; the R. Bravo, 109; the RES, 134; the MR, 59, 61, 85, 92; in the narratives of the De Soto expedition, 57 ff; on Minet's map, 108
Granja, the marqués de la, describes the Gulf of Mexico, 126-127
Great Lakes, 97, 116, 121, 123
Green Bay, xi, 123
Griffon, the, 116
Guacacoya, Guachoya, 85
Guacane, Guancane, Lacane, 86, 89
Guadalquibil, Guadalquibir, 26, 27, 35
Guasuli, Guaxule, 59, 85
Gunthersville, Ala., 60, 64
Gutiérrez, D., 41, maps, 30, 42

H

Haiti, 139
Harleian map, the, 31
Harrisse, H., 21, 26
Havana, 58, 59, 115, 118
Hennepin, L., *Description de la Louisiane,* 135; map in, 123
Henry II map, the, 31
Herrera y Tordesillas, A. de, 81, 83, 91; *Descripcion de las Indias,* 82, 85; map, 85; narrative of the De Soto expedition, 93
Heylin, P., map, 79
Hirrihigua, 86
Hitanichi, R., Apalachicola R., 142
Hiwassee R., 59, 60
Hodge, F. W., 53
Hoeius, F., map, 79
Homem, D., map, 6
Homem, L., map, 3-4
Honda, B., Tampa B., 5, 61, 64
Hondius, J., 80, *Nouveau Theâtre du Monde,* 7; maps, 6, 78, 79
Honduras, Gulf of, 15
Hudson Bay, 100
Huygen, Ian, 81, 84, 85

I

Iberville, P. Le Moyne d', at Cap Français, 139; expedition to the Gulf of Mexico, 135, 137-138; enters the MR, 141; maps of the Gulf of Mexico, 140; questions to, 144
Iguas, 40
Iroquois, 98, 99, 102
Illinois R., xi, 44, 102
Isabella of Portugal, 21, 22
Isabelle, I., 11

J

Jaillot, H., 8, 92, note 23
Jamaica, 12
James Bay, 35
Japan, 97
Jesuit Relations, the MR in the, 97-99
Jolliet, L., xi, 96; expedition of 1673, 100-101

INDEX

Joly, the, 107
Jordán de Reina, J., 138; log, 118-119, 131; site of the BES according to, 120
Juan Ponce, B. de, 5, 25, 32, 37; R. de, 37
"The Jump", 14
Junta de Guerra, 115, 127, 130

K

Kankakee, R., 122
Keys, see Florida Keys
Kohl, J. G., 39, 40

L

La Barre, J. A. Lefebre de, the MR is the RES, 108-109, 141
Labrador, 20, 24
Lacane, Guacane, Guancane, 86, 89
Laclede Co., Mo., 76
Laoton, Lacoto, 15, 34, 41
La Salle, N. de, 138
La Salle, R. Cavelier de, 96, 99, 102, 134, 135, 136, 141; at Matagorda Bay, 115; colony, 118, 125, 127, 138; site, 124, 128; at BES, 120, 123; landing place, 117, 121; map, 8, 102, 106, 107; voyage of 1682, 110, 113, 121; voyages, 116, 122, 132
Le Clercq, C., *Premier Etablissement de la Foy,* 137
Léogane, 139
León, A. de, 125, 130
Little Tennessee R., 60
Lodo, C. de, 118, 139
Lodo, L. de, 139; identified as the BES, 143, 144; as the Mississippi Sound, 143; as Lake Pontchartrain, 144
López de Cerrato, A., 56, 62
López de Gomara, F., 7, 81, 84, 85; describes the American Coast, 48; the Gulf of Mexico, 2-3, 6
López de Velasco, J., 64, 82, 84; describes Florida, 65; the Gulf of Mexico, 5; mentions the maps of Santa Cruz, 73

Louis XIV, 134
Lowery, W., 68
Luna y Arrellano, T. de, expedition, 65

M

Madalena, Magdalena, R. de la, (in Texas), 26, 110
Magdalena, R. de la, (in Florida), 53
Magellan, Strait of, 20, 24
Maggiolo, V. de, map, 15, 16
Malabrigo, 26
Maldonado, F., to Achuse, 58, 73; to RES, 58, 61
Maps, groups of, 9; of the Gulf, 4; of the New World in the sixteenth and seventeenth centuries, xi-xii
Mar Vermejo, Gulf of California, Vermilion Sea, 97
Marin, the, 139
Marly, globe of, 111
Mar pequeña, 1, 4, 26, 27, 30, 64, 76, 89, 125; name of the BES, 2-3
Marquette, J., xi, 99, 100
Martínez, F., 140
Martínez, J., map, 44-45
Matagorda B., 107, 124; B. St. Louis, 137, B. San Bernardo, 123; BES, 125-126, 130, 143; La Salle at, 115
Matal, J., map, 6, 76
Mazza, G. B., map, 77
Medam, *i.e.,* Medanos, R. de, 27
Membré, Z., 110
Mer du Sud, 97; see South Sea, Pacific Ocean
Mercator, G., 3, 80; map of America in *Atlas,* 79; globe of 1541, 26-27, 35, 42; planisphere of 1569, 39-43, 45, 74, 77, 78, 87, 88
Mercator, M., map, 44
Mercator, R., map, 44
Mercator-Hondius, *Atlas,* 79
Merian, M., map, 79
Mexico, 40, 100, 105, 127
Mexico, Gulf of, 5, 65, 97, 98, 99, 101, 109, 113, 115, 127, 128; see Gulf of New Spain, Sea of Florida;

earliest map of, 12; maps of, 4, 29, 31-32, 105, 138, 140, 143; on the padrón, 22, 57; coast, account of voyages along the, xi, 47 ff; description by López de Gomara, 2-3, 6; López de Velasco, 5; by Sigüenza y Góngora, 126, 130-132; on the maps of Berteli, 37; of De Laet, 7, 82-83; on the Paris Gilt Globe, 34; on the Ulpius Globe, 35; on the map of Zaltieri, 37; nomenclature of the coast, 27, 32, 34; on Gastaldi's map, 36; on Gutíerrez' map; on Martínez' map, 45; on Mercator's planisphere, 42; rediscovery of the coast, 113 ff; explored by Barroto-Romero, 118 ff, 126 ff
Mexico City, 36, 66, 117, 118, 127, 128
Michigan, L., xi, 123
Michilimackinac, xi, 103
Micipipi, Misipipi, B., 115, 117, 119, 122
Minet, —, 138, map, 102, 107-108
Miruelo, D., 50, 51
Miruelo, Mirguelo, Vulguelo, B. de, 5, 25, 27, 31, 47, 83
Mississippi R., xi, xii, xiii; the Chucagua, 85, 92, 94, 138; not the Chucagua, 104; the Colbert R., 105, 111, the R. Escondido, 104, 105, 111, 136, 137; very probably the RES, 101; is the RES, 108-109; the MR is not the RES, 146; called R. Grande, 57, 59, 61, 85, 92; was thought to be the Ohio, 99, 100; is Oquechiton, 68; named R. de la Palizada, 113, 118, 132, 138, 140, 141; the Jolliet-Marquette expedition, 100; La Salle's descent of the, 116; mouths of the MR, 101, 107, 110-11, 136, 141; description of the "MR", 13; description of the "MR" by Cabeza de Vaca, 50, 51-52, 54; description of the MR by the chroniclers of the De Soto expedition, 119; by Galinée, 100; in the *Jesuit Relations,* 97-99; in La Salle's writings, 103-107; the MR on the map of Bernou-Peronel, 100; of Coronelli, 102, 111; of Delisle, 137; of Franquelin, 102, 107; of La Salle, 106; of Minet, 102
Mississippi Sound, 140, 143, see L. de Lodo
Mississippi, State, 11
Mississippi Valley, xii, 10, 93
Missouri R., 16, 70, 83, 102
Mobile B., 51, 59, 114, 118, 126, 131, 132, 134; identified as the BES, xiii, 123, 126, 128-129, 130, 145
Mobile R., xii
Mocoço, Mocosa, 40
Moctezuma, Conde de, 134
Monclova, Conde de, 123
Montañas, R. de, de las, 25, 63, 83
Montañas altas, R. de, 34
Montezuma, 14
Montreal, 96
Morales, A. de, map, 19
Moscoso, L. de, xiii
Mouila, 114

N

Naguater, Naguatex, Neguateix, Neguater, 76, 86, 89
Napissa, Napochies, 60
Napochies, Natchez, Napissa, 66, 67, 68, 69
Narváez, P. de, 47, 55; expedition, 49, 84, landing place, 50
Natchez, 68; see Napochies
Navidad, 27
Negrillo reefs, 127, 128
Neguateix, 76; Naguater
New Biscay, 94
New France, 96, 108, 115, 116, 135
New Granada, 97; see Nouvelle Grenade
New Mexico, 94, 114, 128
New Spain, 71
New Spain, Gulf of, 5, 65, 127, 128; see Gulf of Mexico
New Sweden, 100
Nieves, R. de, 25, 27, 83, 84, 87

INDEX

Nolin, J. B., 135, 136
Norte, R. de, 97
Nouvelle Grenade, 97; see New Granada
North Sea, 71, see Atlantic Ocean, Sea of the North
Nuestra Señora de la Concepción y San Joseph, the, 118

O

Ochus, 73, see Achuse
Ochus, R. de Santa Maria de, 90
Ogilby, J., map, 79
Ohio, R., 28, 44, 60, 74, 83; thought to be the MR, 99, 100
Olibahali, 66, see Ulibahali
Ontario, L., 99
Oquechiton, R., identified as the RES, 67, 68; as the MR, 68; neither the RES nor the MR, 68, 69
Oro, R. del, 45, 64
Ortelius, A., 3, 6, 7, 76; maps, 43, 77, 88; *Theatrum,* 72
Otagil, 40, 41; see Tagil
Otero Co., New Mex., 76
Ottawa R., 99
Oviedo y Valdes, G. F. de, 1, 48, 53, 73; describes Chaves' map of 1536, 2-3; Ranjel's narrative of the De Soto expedition in, 56, 57, 59, 62; comments on Cabeza de Vaca's narrative, 49; on the *Padrón General,* 24-25

P

Pacaha, 89
Pacific Ocean, Sea of the South, South Sea, 70, 97
Pacoa, 64
Padilla, J. de, 71
Padrón, the, 4, 16 ff, 42, 46, 51, 57, 63, 73
Padrón General, 19, 20, 21, 22, 23, 24-25, 29, 30, 32
Padrón Real, 18, 19, 20, 22, 23
Pafalaya, 63
Palacios, G. de, 115-116, 137; site of the BES, 121; La Salle's landing place, 117; reports on the Barroto-Romero expedition, 120
Palizada, Empalizada, R. de la, R. de la Palissade, des Palissades, 120, 121, 123, 134, 135; name of the MR, 118, 132, 138, 140, 141; location, 119, 139
Palmas, R. de las, R. la Palma, Palmar, R. des Palmes, 14, 25, 26, 34, 43, 50, 51, 84, 111
Pánuco, 5, 15, 30, 50, 53, 54, 71, 94, 104, 106, 140; R. de, 12, 14, 15, 17, 34, 41, 42, 111
Paris Gilt Globe, the, 34, 35
Park Co. Colo., 76
Patoulet, —, 138
Paz, R. de la, 25
Peñalosa, D. de, 109
Pensacola B., 51, 52, 53, 59, 118, 124, 128, 140; Achussi, 130; Santa Maria de Galve, 134; map of, 131; reports on, 134; occupation by the Spaniards, 126 ff
Perla, R. de, 42
Peronel, C., 110
Pescadores, Piscadores, R. de, R. des pescheurs, 3, 25, 26, 30, 43, 64, 106
Petit Goave, 115, 140
Pez, A. de, 132; memorial, 126 ff; Laguna de, 139
Piloto mayor, 18, 19
Pineda map, the, 1, 12-13, 14, 15, 16
Plancius, P., maps, 77, 79
Plancius-Vrient map, the, 79
Ponce de León, J., 13, 85
Pontchartrain, L. Phélipeaux de, 138, 139, 140, 142
Pontchartrain, L., 144
Pontotoc Co., Miss., 60
Portage, Wis., xi
Port Nelson, 97
Prairie du Chien, Wis., xi

Q

Quebec, 96
Quiguate, 89

Quivira, 70, 71, 94
Quizquiz, 59, 62

R

Ranjel, R., 40, 70, 72, 73; narrative of the De Soto expedition, 55, 59, 61, 62, 91, 93; report in Oviedo, 56, 59; voyage of Maldonado, 58
Red, R., 102
Relation officielle of La Salle's descent of the MR to the sea, 110
Ribero, D., 19, 21, 23, 29; author of the Weimar map (?), 22; maps, 21, 22, 25, 26, 27
Richelet, P., adapts Garcilaso de la Vega's narrative, 92, 103
Rocky Mountains, xii
Romero, A., 118, 119, 121, 123, 125
Romo, C., 30
Rotz, J., atlas, 31, note 42
Ruysch, J., map, 11
Ryswick, Treaty of, 134

S

Sabine, L., the BES (?), 145
Sabine R., 114, 121
St. Clement's Point, 50
St. Francis R., xiii, 60, 89
St. Francis Xavier mission, 98
St. Ignace, Mich., xi
St. Joseph R., 122
St. Lawrence R., 40, 99, 100, 116, 123; see Canada R.
St. Louis, B., Matagorda B., 137
St. Louis, B., Miss., 144
St. Mark's, B., 47, 51, 52, 114, 120; R., 53
Salvador, Matas, Montas de, 27, 30
Salviati map, the, 1, 16-17, 22
San Bernardo, Bernardino, Matagorda B., 123, 128, 129, 130, 134, 140
San Blas, C., 25, 51, 90
San Diego Is., 141
Sanson, A., 93
Sanson, G., 93
Sanson, N., 6, 96, 101; map of 1650, 7, 87-88, 97, 98; map of 1656, 7, 87, 88-91, 94

Sanson-Jaillot map, the, 8, 92-94, 103, 108, 110
Santa Barbara Mountains, 103
Santa Cruz, A. de, 24, 29; author of the De Soto map, 62; map of 1542, 27-28, 71, 72, 74; other maps, 2, 5, 24, 30, 65, 73
Santa Maria de Galve, B., 134; see Pensacola B.
Santo Domingo, 56, 116, 117, 137, 140
Santo Domingo, Audiencia de, 47, 48, 49, 62
Scaife, W. B., xi-xiii, 125, note 18
Seco, R., 75
Shea, J. G., 57, 66, 68
Ship I., 140, 143
Sigüenza y Góngora, C. de, 129, 131, 132, 134, 135; describes the Gulf Coast, 130-132; writings, 126 ff
Solo, R., 25, 43
Sonora, 47
South Carolina, 11
South Sea, Sea of the South, Pacific Ocean, ix, 70, 97, 98
Southern United States, xi, 55, 72, 82, 98, 119; map of, 77; map by Chaves, 74-76; by Mercator, 39-40; by Santa Cruz, 73, 74-76; see Florida
Spagnola, I., 11
Steck, F. B., 68
Suala Mountain, 40, 88; see Xuala
Superior, L., 97
Suwannee R., 53, 83

T

Tacobaga, Tacobago, Tocobaga, 5, 83, 84, 90
Taensa portage, 106
Tagil, Otagil, 40, 41
Talladega Co., Ala., 66, 69
Tallapoosa R., 66
Talissi, 63
Tamacho, 14, 34
Tampa B., 5, 50, 61, 64, 83, 144
Tampico, R. de, 9, 29, 113, 117, 120, 121; see Pánuco

INDEX

Tascalusa, 63, 86
Tatton, M., map, 78
Tennessee R., 59, 60, 64
Tesa, Costa, 26
Texas Coast, 41, 47, 64, 74, 102, 105, 111
Theatrum of Ortelius, 72
Thévenot, M., map, 123; location of the mouth of the MR, 136
Thevet, A., map, 43-44
Thomas, D., 115
Tiachi, 63
Tiguex R., 69
Todos Sanctos, B. de, 36
Tombigbee R., 114
Tonti, H., 103, 106, 109, 110
Trinity R., 132; the RES (?), 145
Tronson, L., 106
Tropic of Cancer, 16, 39
Tula, 89
Turin map, the, 22, 34

U

Ulibahali, Ullibahali, 66; see Olibahali
Ulpius globe, the, 35
Utiamgue, 89; see Autiamque

V

Van den Keere, P., map, 79
Velasco, L. de, 65-66
Velasco I., 52
Velasquez, D., 13, 15
Vélez, marqués de los, describes the Gulf of Mexico, 127-128
Vera Cruz, 15, 113, 114, 115, 117, 118, 121, 123, 127, 128, 129, 130

Verde, R., 32
Vermilion Sea, Gulf of California, Mar Vermejo, 97, 98, 99, 100, 101
Verrazano, H. de, map, 15-16
Vicente, J., 140
Villiers du Terrage, M. de, Galveston Bay is the BES, 132, note 15, 145
Virginia, 98, 100, 101, 107
Vopel, G., map, 35-36
Vulguelo, Miruelo, B. de, 27

W

Waldseemüler, M., map, 11
Washington, D. C., 16
Weimar map, the, 1, 20-21, 22, 25, 26, 47
Wilson Co., Kansas, 76
Winsor, J., 39, 40, 57
Wisconsin R., xi
Withlacoochee R., 53
Wolfenbüttel map, the, 1, 21, 22, 26, 27
Wytfliet, C., 81; map, 6, 76, 78

X

Xuala, Xualla, 40, 56, 59, 62, 88; see Suala
Xuárez, J., 50, 51

Y

Ychiaha, 89; see Chiacha
Yucatan peninsula, 12, 15, 127

Z

Zaltieri, —, map, 37, 39, 41
Zierto, d., *i.e.*, Desierto, C., 27